Inversions

Popular Lectures in Mathematics

Survey of Recent East European Mathematical Literature

A project conducted by
Izaak Wirszup,
Department of Mathematics,
the University of Chicago,
under a grant from the
National Science Foundation

I. Ya.
Bakel'man

Inversions

Translated and
adapted from the
Russian edition by
Joan W. Teller and
Susan Williams

The
University of Chicago
Press
Chicago and
London

The University of Chicago Press, Chicago 60637
The University of Chicago Press, Ltd., London

International Standard Book Number: 0–226–03499–2
Library of Congress Catalog Card Number: 74–5727

Contents

Preface

In the study of plane geometry, various *transformations* of geometric figures often play an important part. Of these transformations, the so-called isometries and dilations are most commonly discussed in elementary treatments. An important property of these transformations is that they preserve basic geometric classifications: Straight lines "go into" straight lines and circles "go into" circles. *Inversions* are more complicated transformations of geometric figures, under which straight lines may be mapped to circles, and conversely. The use of such mappings allows us to develop a unified method of solution for many of the problems of elementary geometry, especially those concerning constructions and *pencils* of curves. The result is that the theory of inversions lends a less artificial character to the interrelationships among types of geometric figures. The approach used in this theory is also useful in *boundary* questions arising in elementary and "higher" geometry. It also enables us to provide an interpretation of the so-called Lobachevskian geometry in the Euclidean plane. There are interesting connections between inversions and the complex numbers or, more accurately, elementary functions whose range and domain are the complex numbers.

This book discusses the inversion transformations and their applications. To provide the most convenient presentation possible, the material is divided into three chapters.

In the first chapter, we shall study inversion transformations and their applications to questions in elementary geometry. In the second chapter, it will be shown that the transformations of the first chapter can be expressed as linear and *linear-fractional* functions of a complex variable. We shall also establish that, conversely, each such function defines a transformation of the plane which reduces to a sequence of isometries and inversions. In the third chapter the foundations of geometry are presented from the standpoint of group theory; using these foundations and relying on the material in chapters 1 and 2, we briefly construct Euclidean plane geometry and Lobachevskian plane geometry.

The reader can find a more detailed presentation of the material touched upon here in chapter 3 of *Vysshaya geometriya* [Higher geometry] by N. V. Efimov.

This book is based on lectures given by the author at various times to students in Leningrad.

1

Inversions and Pencils of Circles

1.1. Elementary Transformations of the Plane

The idea of transforming one geometric figure into another will play a fundamental role in this book. In this section we shall discuss figures in the plane. But first of all, we wish to state precisely what we mean by *transformations* of geometric figures. Consider a plane, and let us assume that we have some rule that, for each point X in the plane, determines a corresponding point X' in the same plane. This rule of correspondence (let us call it T) is called a *transformation* of the plane, and the point X', corresponding to the point X, is called the *image* of X under T. Transformations of the plane will be written in capital letters. If T is some transformation of the plane, and if X is some point in the plane with image X' under T, we write $X' = T(X)$.

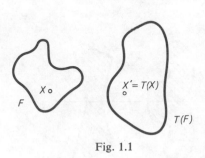

Fig. 1.1

Suppose we are given a transformation T of the plane and a plane figure (for example, a line or a circle) F. T takes each point X of the figure F into some point X', its image. The figure F', consisting of all the points which are images of points in F, is called the *image of the figure F* under the transformation T. We shall often denote the figure F' by $T(F)$ (see fig. 1.1).

Usually, a point and its image do not coincide. When the point X and its image $T(X)$ do coincide, the point X is called a *fixed point* of the transformation T.

The transformation of the plane taking each point X into itself is called the *identity transformation*. In other words, a transformation of

1

the plane is the identity if and only if all the points of the plane are fixed points. We shall denote the identity transformation by the letter I.

A plane figure F is called *invariant* under a transformation T of the plane if the image of F coincides with F, that is, if

$$F = T(F).$$

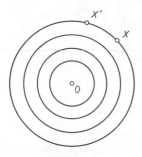

Fig. 1.2

It is important to note that a figure invariant under a transformation need not have a single fixed point under that transformation. For example, if T is a rotation of the plane through some fixed nonzero angle about a point O, then the only fixed point of T is O.[1] Thus all non-degenerate circles with center O are invariant under T, and yet none of them contains a single fixed point (fig. 1.2).

We shall now examine the elementary transformations of the plane in greater detail.

1.1.1. *Reflection with respect to a line.* We define the *reflection of the plane with respect to the line l* by the following rule: If a point X lies on l, it is carried into itself. If the point X does not lie on l, then we take as the image of X the point X' that is symmetric to X with respect to the line l (fig. 1.3).

The figures invariant under reflection with respect to the line l are all those figures which have the line l as an axis of symmetry, including l itself. Two such invariant figures are shown in figure 1.4.

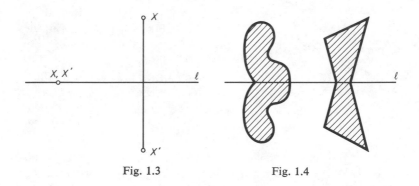

Fig. 1.3   Fig. 1.4

1. By a *nonzero angle* we shall mean an angle whose radian measure is not an integral multiple of 2π.

All points of the line *l*, and only those points, are fixed points of the transformation.

1.1.2. *Parallel translation.* A *parallel translation* of the plane is defined by the following rule: Suppose we are given a segment *AB* of the line *l* in the plane; if the point *X* does not lie on the line *l*, then its image *X'* is the fourth vertex of the parallelogram constructed with sides *AB* and *AX*. If *X* lies on the line *l*, then for *X'* we take the point of *l* such that the line segments *AX* and *BX'* are of equal length and the line segment *XX'* has the same length as the line segment *AB*. In this way the parallel translation translates each point of the plane by the distance *AB* in the direction moving from *A* to *B* (fig. 1.5). In terms of vectors, each point of the plane is translated by the vector **AB**; that is, for each point *X* in the plane, the vector equality **XX'** = **AB** holds (fig. 1.6).

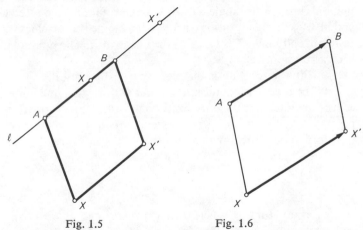

Fig. 1.5   Fig. 1.6

If the vector **AB** is the zero vector (that is, if the point *A* coincides with the point *B*), then the parallel translation by the vector **AB** is the identity transformation.

Let *T* be a parallel translation by a nonzero vector **AB**. It is obvious that *T* has no fixed points. Figures invariant under *T* include, for example, all lines parallel to the line determined by the segment *AB*. There are many other invariant figures; figures 1.7 and 1.8 depict figures *L* and *Q* which are invariant under *T*. The curves L_k and Q_k are the images of the curves L_{k-1} and Q_{k-1} respectively.

1.1.3. *Rotation about a point.* Let *O* be a given point in the plane and *α* (read "alpha") a given angle. We define the *rotation* of the plane through the angle *α* about the point *O* by the following rule: If *X* is an

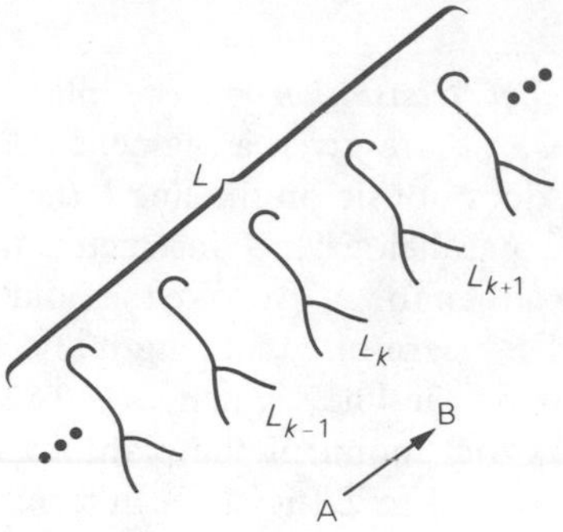

Fig. 1.7 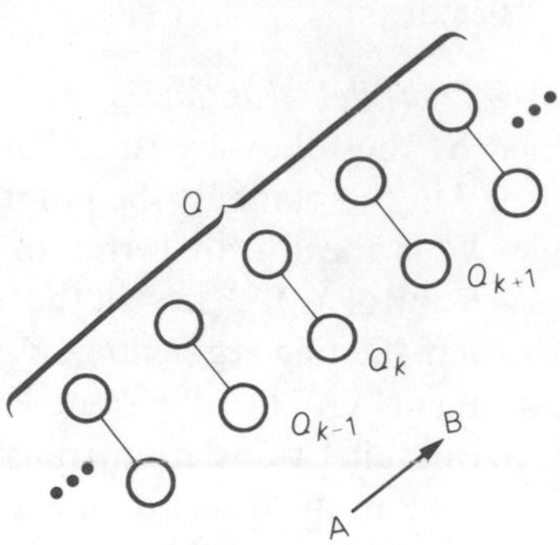   Fig. 1.8

arbitrary point in the plane, we rotate the line segment OX about the point O through the angle α (if $\alpha > 0$, the rotation is counterclockwise, and if $\alpha < 0$, the rotation through an angle $|\alpha|$ is clockwise). The resultant endpoint X' is taken as the image of X. The point O is fixed in such a rotation.

If $\alpha = 0$, the rotation is the identity transformation.

Let T be the rotation about the point O through some nonzero angle α. It is obvious that the only fixed point of the transformation T is the point O. Circles having O as their center are invariant figures under this transformation. If the angle α has the radian measure

$$\alpha = \frac{2\pi}{n},$$

where n is a natural number, then a regular m-gon inscribed in a circle with center O is invariant under T if and only if the number of sides m is divisible by n (fig. 1.9). In figure 1.10 we see a more complicated invariant figure.

1.1.4. *Isometry.* An *isometry* is a transformation of the plane which "preserves" distances between points. That is, T is an isometry if and only if for any pair X and Y of arbitrary points in the plane, the line segments XY and $T(X)T(Y)$ are of equal length (or, equivalently, the distances XY and $T(X)T(Y)$ are equal). We require further that the transformation T be *one to one* and *onto*; that is, that every point in the plane be the image of some other point (T is onto) and that no two distinct points have the same image. It is easy to see that all of the transformations described above are isometries. In a certain sense the converse is true: It can be shown that any isometry is either a rotation, a

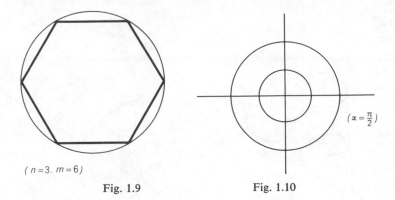

(n = 3, m = 6)

$(\alpha = \frac{\pi}{2})$

Fig. 1.9   Fig. 1.10

parallel translation, a reflection with respect to a line, or some com-position (successive application) of these.

1.1.5. *Dilation.* Let us fix some point O in the plane, and let $k > 0$ be some fixed number. The *dilation with center O and coefficient k* is the transformation of the plane which takes the point O into itself, and takes any point X different from O into the point X' lying on the ray (half-line) OX and satisfying

$$OX' = k \cdot OX.$$

If $k = 1$, then the dilation is the identity transformation. If $k \neq 1$, then the only fixed point of the transformation is the center of the dilation, the point O. We note that if $k < 1$, a given figure "shrinks" under dilation, while for $k > 1$, it "expands." Rays having their initial points at the center of the dilation are clearly invariant under dilation.

It is possible to exhibit, in a fairly simple way, a more complicated invariant figure. Let F be some figure in the plane (fig. 1.11). We denote by mF the figure F' which is the image of F under the dilation with center O and coefficient m. Given the dilation T with coefficient k and center O, we consider figures

$$\ldots, \frac{1}{k^m} F, \frac{1}{k^{m-1}} F, \ldots, \frac{1}{k} F, F, kF, \ldots, k^{m-1}F, k^mF, \ldots.$$

The figure G, representing the union of all these figures (fig. 1.12), as is easily shown, is invariant under the transformation T.

Finally, let us make use of the concepts of isometry and dilation to formulate precise and general definitions for the terms *congruent* and *similar*, which play an important role in elementary geometry:

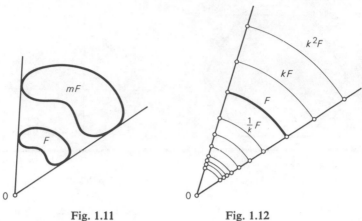

Fig. 1.11   Fig. 1.12

Two figures F_1 and F_2 are said to be *congruent* if there exists an isometry taking the figure F_1 into F_2. The figures F_1 and F_2 are said to be *similar* if there exists a dilation taking the figure F_1 into some figure F_2' which is congruent to the figure F_2.

1.2. Stereographic Projection: The Point at Infinity of a Plane

The concept of a transformation, considered in §1.1 for the plane, clearly extends to any geometric figure (including subsets of the plane and of three-dimensional Euclidean space). If the image of the figure M under such a transformation T covers the entire figure N, we say that T is a transformation of M *onto* N.

In the study of the inversion transformations, it is quite useful to examine one particular transformation of the three-dimensional sphere onto the plane. This transformation is called the *stereographic projection* and is defined as follows: Let K be a sphere and P a plane tangent to K at a point S (fig. 1.13). The point S will be called the *south pole* of K, and the diametrically opposite point N, the *north pole*. Let X be any point of K other than N. Then the point X' at which the ray NX intersects the plane P is taken to be the image of X. Clearly, the entire plane P is covered. Thus, the stereographic projection transforms the sphere K, minus the point N, onto the entire plane P.

Let us consider how the image of the point X on the plane P changes as X approaches the point N. From the similar right triangles $X'NS$ and SNX (fig. 1.14), we have

$$\frac{SX'}{NS} = \frac{XS}{NX}.$$

Hence,

$$SX' = \frac{NS \cdot XS}{NX}.$$

Let r be the radius of the sphere K. Then, for a point X sufficiently close to the north pole N, $XS > r$, and therefore

$$SX' > \frac{2r^2}{NX}$$

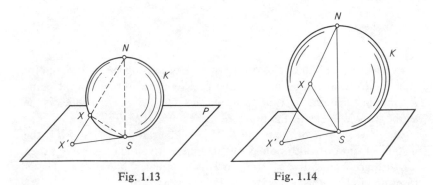

Fig. 1.13   Fig. 1.14

(since $NS = 2r$). It is obvious that as the point X gets arbitrarily close to the point N (NX approaches zero), the length of the line segment SX' increases without bound, so that the point X' gets unboundedly further away from the point S. Consequently, the point N cannot go into any point of the plane P under the stereographic projection. In order to extend the stereographic projection to the entire sphere K, that is, for the north pole N to be given an image in the plane P, we must add a new point to P. The added point O_∞ is called the *point at infinity*. Now, letting the north pole N go into the point at infinity O_∞, we find that the stereographic projection takes the sphere K onto the plane P.

Let us consider some of the properties of the point at infinity. Let l' be any straight line in the plane P. We consider the plane through the point N and the line l' (fig. 1.15). This plane intersects the sphere K in some circle l passing through the point N. The line l' is obviously the image of the circle l under the stereographic projection. On the other hand, the image of every circle on the sphere K passing through the point N is represented by a line in the plane P which is the intersection of the plane determined by the circle l, and the plane P. It follows that the stereographic projection establishes a one-to-one correspondence

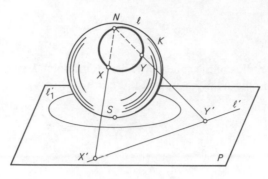

Fig. 1.15

between the set of all circles on the sphere K passing through the point N and the set of all lines in the plane P. Therefore, any line in the plane P contains the point O_∞ (and thus all lines intersect at O_∞), which is the stereographic image of the point N.

Let l_1' be a circle in the plane P. If r' is the radius of l_1' and d is the distance from the south pole S of the sphere K to the center of l_1', then the distance from S to any point of l_1' is no greater than $d + r'$. Therefore, no circle in the plane P contains the point at infinity.

As we know, any three noncollinear points determine a circle. Lines in the plane are similarly determined by three points, two of which may be chosen arbitrarily and the third of which is the point at infinity. Therefore, a line may in a certain sense be regarded as a circle having as one of its determining points the point at infinity.

Now consider the set of all circles on the sphere K whose planes are parallel to the plane P. We shall consider this set to contain the points S and N as degenerate circles of zero radius. The stereographic projection of this set of circles (fig. 1.16) is the set of all concentric circles in the plane P with center S, which includes the point S (fixed by the stereographic projection) and the point at infinity (the stereographic image of the point N). Since the point of tangency of the sphere K and the plane P could be any point of P (simply make the proper translation of the sphere K parallel to the plane P), we can consider any system of concentric circles to contain the common center of all the circles and the point at infinity.

1.3. Inversions

Let us fix a circle in the plane P with center O and radius r. An *inversion of the plane with central point O and radius r* is the transforma-

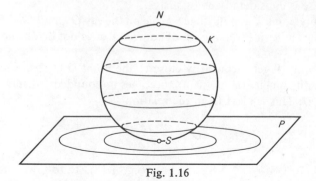

Fig. 1.16

tion of the plane determined by the following rule: A point X different from the points O and O_∞ is carried into the point X' on the ray OX which satisfies the equation

$$OX' = \frac{r^2}{OX}$$

(fig. 1.17); the point O is taken into the point O_∞, and the point O_∞ is taken into the point O.

The circle depicted in figure 1.17, with radius r and center O, is called the *circle of inversion*. If X lies on the circle of inversion, then $OX = r$ and, consequently,

$$OX' = \frac{r^2}{OX} = r .$$

Since the points X and X' both lie on the ray OX, the points X and X' coincide. It follows that all the points on the circle of inversion are fixed points and that the circle of inversion itself is an invariant figure.

A point different from O lying inside the circle of inversion is taken by the inversion to a point lying outside the circle, and, conversely, a point different from O_∞ lying outside the circle of inversion is carried to a point in the interior of the circle.

In the first case, we have $OX < r$, and thus

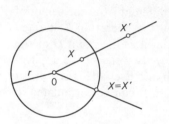

Fig. 1.17

$$OX' = \frac{r^2}{OX} > \frac{r^2}{r} = r ,$$

verifying that the point X' lies outside the circle of inversion. The second case is considered analogously.

Thus, any point X and its image X' lie on the ray OX and on different sides of the circle of inversion, if, of course, X does not lie on the circle (fig. 1.17).

If the point X gets arbitrarily close to the point O (OX approaches zero), then its image, the point X', becomes unboundedly distant from the point O. This is clear from the relation

$$OX' = \frac{r^2}{OX}$$

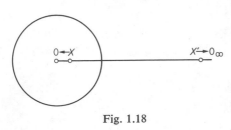

Fig. 1.18

(fig. 1.18). It follows that the point X' approaches the point at infinity. Analogously, we can show that if a point X is made arbitrarily distant from the point O, its image X' becomes arbitrarily close to the point O. Thus the definition of the inversion, which determines O_∞ as the image of O and conversely, is a natural one.

Let X be a point different from O and O_∞ and let T be the inversion of the plane with central point O and radius r. We denote $T(X)$ by X' and $T(X')$ by X''. Then all the points X, X', X'' lie on the same ray OX and satisfy the equations

$$OX' = \frac{r^2}{OX}; \qquad OX'' = \frac{r^2}{OX'}.$$

It follows that

$$OX'' = \frac{r^2 \cdot OX}{r^2} = OX.$$

Thus, if X is an arbitrary point of the plane different from the center of the inversion and the point at infinity, then the operation T iterated twice takes the point X into itself. If the point X is the point O or the point at infinity, the result is the same: Under two successive applications of the inversion T, the point X is taken into itself. This is a direct consequence of the definition of the inversion. It can be formulated in the following theorem:

THEOREM 1.1. *A transformation of the plane which is the composition of an inversion with itself is the identity transformation.*

Finally, we remark that if the inversion T takes the point X into the point X', then T also carries the point X' into the point X; that is, X and X' trade places. We recall that reflections with respect to a line have the same property. This is the reason that inversions are sometimes called *reflections with respect to a circle.*

1.4. Properties of Inversions

In this section we shall fix T as the inversion on the plane with center O and radius r.

First we shall prove a simple lemma which plays an important role in the study of the properties of inversions.

LEMMA 1.1. *Suppose the points A and B in the plane are different from each other and from the points O and O_∞, and that the points O, A, and $B are noncollinear. Let $A' = T(A)$ and $B' = T(B)$. Then the triangles OAB and $OB'A'$ are similar, with corresponding parts indicated by the letter orderings OAB and $OB'A'$.*

Proof. The triangles OAB and $OB'A'$ (fig. 1.19) have a common angle, and the sides including the angle are proportional. To show this, we note that since

$$OA \cdot OA' = \frac{OA \cdot r^2}{OA} = r^2 = \frac{OB \cdot r^2}{OB} = OB \cdot OB',$$

we have

$$\frac{OA}{OB} = \frac{OB'}{OA'}.$$

It follows that triangles OAB and $OB'A'$ are similar. However, in similar triangles, equal angles lie opposite proportional sides, so from the ratio

$$\frac{OA}{OB} = \frac{OB'}{OA'}$$

we get equality of the corresponding angles:

$$\angle OAB = \angle OB'A'$$
$$\angle OBA = \angle OA'B',$$

proving that the letter orderings OAB and $OB'A'$ indicate corresponding parts.

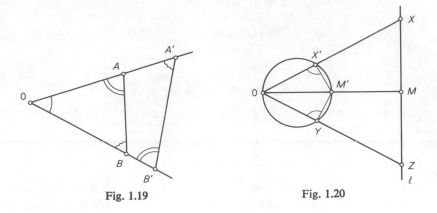

Fig. 1.19   Fig. 1.20

THEOREM 1.2. *The inversion T carries any line passing through the center of inversion into itself; that is, a line passing through the center of the inversion is an invariant figure.*

The proof of this theorem follows easily from the definition of an inversion.

THEOREM 1.3. *The inversion T takes a line not passing through the center of inversion into a circle passing through the point O.*

Proof. Let *l* be a line not passing through the center of inversion *O*. Drop a perpendicular from the point *O* to the line *l*, and let its intersection with *l* be the point *M* (fig. 1.20). Let *M'* be the image of the point *M* under *T*. The point *M'* clearly lies on the ray *OM*. Consider an arbitrary point *X* (different from O_∞) on the line *l*; let *X'* be the image of *X* under *T*. By lemma 1.1, we have

$$\angle OX'M' = \angle OMX = \frac{\pi}{2}.$$

Therefore, by an elementary geometry theorem concerning right triangles and diameters of circles, the point *X'* lies on a circle *K* having the line segment *OM'* as a diameter. Since this statement holds for all points *X* on the line *l*, the image of the line *l* under *T*, *l'*, is contained in the circle *K*.

Now we must prove that the set of points *l'* actually coincides with the set of points of the circle *K*; that is, that *K* is also contained in *l'*. First let us remark that the point *O* is contained in the set *l'*, since *O* is the image of O_∞, which is contained in l. Now let Y be an arbitrary point of the circle K different from O. The ray OY intersects the line l at some point Z; we claim that the point Y is the image of Z under T. Since the points Y and Z lie on the same ray OZ, we need only prove that Y satisfies

$$OY = \frac{r^2}{OZ}.$$

By construction, the triangles OYM' and OMZ (fig. 1.20) are similar. Therefore,

$$\frac{OY}{OM'} = \frac{OM}{OZ}.$$

Hence,

$$OY = \frac{OM \cdot OM'}{OZ} = \frac{r^2}{OZ},$$

the desired result. Thus, Y is the image of Z under T. Since this is true for all Y on the circle K, K is contained in l', and since, by the above, l' is contained in K, we conclude that the image of l coincides with K, the assertion of the theorem.

The constructions carried out in the proof of theorem 1.3 enable us to construct the image of a given line under the inversion T using only a compass and straightedge. From the center of the inversion—the point O—we drop a perpendicular OM (fig. 1.20) to the line l. As before, we construct the point M', which is the image of M (by constructing a line segment of length r^2/OM along the perpendicular). The image of the line l is the circle l' constructed with the line segment OM' as a diameter.

In the special case where the line l is tangent to the circle of inversion, the points M and M' coincide, and the circle l' is constructed with the line segment OM as a diameter. If l intersects the circle of inversion in two points X and Y, then since O is necessarily on the circle $K = l'$, K is completely determined by O and the fixed points X and Y.

THEOREM 1.4. *The inversion T transforms a circle passing through the center of inversion O into a straight line not passing through O.*

The proof follows from the fact that the composition of T with itself is the identity transformation and from theorem 1.3.

THEOREM 1.5. *The inversion T transforms a circle not passing through the center of inversion O into another circle not passing through O.*

Proof. Let K be a circle not passing through O. We construct a line g through the point O so that it intersects the circle K in a diameter AB (fig. 1.21). Let A' and B' be the images of the points A and B under T, X an arbitrary point on the circle K different from A and B, and X' its image.

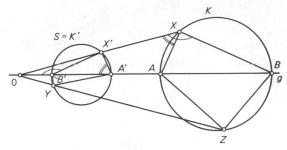

Fig 1.21

By lemma 1.1, the triangles OXA and $OA'X'$ are similar, so that

$$\angle OA'X' = \angle OXA.$$

Analogously, the triangles OXB and $OB'X'$ are similar, and consequently

$$\angle OB'X' = \angle OXB.$$

Since

$$\angle A'X'B' = \angle OB'X' - \angle OA'X' = \angle OXB - \angle OXA$$

$$= \angle AXB = \frac{\pi}{2},$$

it follows that X' lies on a circle S having the line segment $A'B'$ as a diameter. Since X was an arbitrary point of the circle K, the image K' of K under T is contained in the circle S. To show that K' coincides with S, we must prove conversely that S is contained in K'. Let Y be an arbitrary point of the circle S different from A' and B' and Z the point on the ray OY satisfying

$$OZ = \frac{r^2}{OY}.$$

It is obvious that the point Z is taken by the inversion T onto the point Y. Further, since the points A', B', and Y are the images under T of A, B, and Z respectively, lemma 1.1 allows us to conclude that

$$\angle AZB = \angle OZB - \angle OZA = \angle OB'Y - \angle OA'Y = \angle A'YB' = \frac{\pi}{2}.$$

Consequently, the point Z lies on the circle K. It follows that the figures S and K' coincide. By construction, the endpoints of the diameter of the circle K—the points A and B—are different from O and are located on the ray OA. Therefore, the circle K' does not pass through the point O (or, alternatively, if K' were to pass through O, K would have to pass through O_∞; yet no circle contains O_∞).

The constructions performed above enable us to construct the image of a circle under an inversion with compass and straightedge. Let us consider this question in greater detail.

Case A. The circle K does not pass through the center of inversion. In this case, we construct a ray from the point O which intersects the circle K in a diameter AB. We then construct A' and B', the images of the points A and B respectively. The circle K', the image of the circle K under the inversion T, is just the circle with the line segment $A'B'$ as a diameter (fig. 1.22).

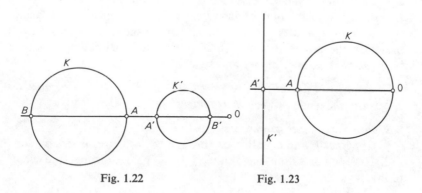

Fig. 1.22   Fig. 1.23

Case B. The circle K passes through the center of the inversion. In this case, by theorem 1.4, the image of K is a line K'. We construct the ray OA from the point O (fig. 1.23) which intersects K in the diameter OA. We then construct the image of A, A'. The line perpendicular to the ray OA at the point A' is the desired line K'.

The construction of the line K' can be significantly simplified in two instances:

1. If the circle K intersects the circle of inversion in two points B and C, then the line K' coincides with the line determined by segment BC (fig. 1.24).

2. If K is tangent to the circle of inversion at some point, then the line K' is tangent to the circle of inversion at that same point (fig. 1.25).

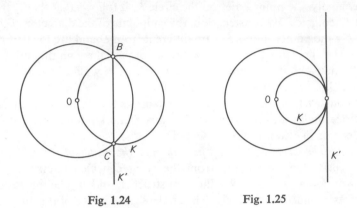

Fig. 1.24   Fig. 1.25

We shall now consider how angles between curves are affected by the operation of an inversion T. As we know, the angle between two curves L_1 and L_2 at their point of intersection is the smaller of the angles between their tangents at that point. It can be shown that an inversion preserves the angles between curves. We shall prove this statement below for the cases of circles and straight lines.

THEOREM 1.6. *Under an inversion T, the angle between straight lines is equal to the angle between their images.*

Proof. Three cases can be presented here:

1. The lines l_1 and l_2 both pass through the center of inversion O.

2. Exactly one of the lines, either l_1 or l_2, passes through the center of inversion O.

3. Neither l_1 nor l_2 passes through the center of inversion O.

In the first case the theorem is obvious. Let us consider cases (2) and (3). In (2) (fig. 1.26) we assume, without loss of generality, that the line l_1 passes through the center of inversion O and that the line l_2 does not. Then the inversion T takes the line l_1 into itself; that is, the image of l_1 coincides with l_1. The line l_2 does not pass through the center of the inversion and, therefore, is taken by the inversion into a circle l'_2 passing

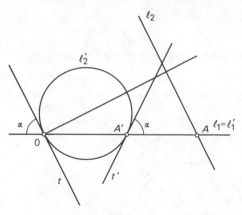

Fig. 1.26

through the point O. The tangent t to the circle l_2' at the point O is parallel to the line l_2 (fig. 1.26).

With respect to the relative position of the lines l_1 and l_2, there are two possibilities:

a. the lines l_1 and l_2 can be parallel;

b. l_1 and l_2 can intersect at a point A.

If l_1 and l_2 are parallel, the angle between them is clearly zero. But the line l_1 passes through the point O and is parallel to l_2. Therefore, it must coincide with the tangent t to the circle l_2' at the point O. It follows that the angle between l_1' and l_2' is equal to zero, and consequently the theorem is proved for the case (*a*).

Now let l_1 and l_2 be nonparallel, with A their point of intersection. Let α be the smaller of the angles between $l_1 = l_1'$ and the line l_2, which is equal to the angle between l_1 and the line t. The point A is taken by the inversion into some point A' which is the intersection of the line l_1' and the circle l_2'. But the line $l_1' =$ the line OA' must intersect the tangent t' to the circle l_2' at A' at the same angle at which it intersects the tangent t to l_2' at O. Since t is parallel to l_2, this angle is α, and the proof for case (2) is complete.

The third case (fig. 1.27) may be proved analogously. We remark only that if the lines l_1 and l_2 are parallel, the corresponding circles l_1' and l_2' are tangent at the point O and thus intersect in a zero angle, the same angle as is formed by the parallel lines l_1 and l_2. If the lines l_1 and l_2 intersect, then, as is evident from figure 1.27, the angle between the circles l_1' and l_2' at the point O is equal to the angle between the lines l_1

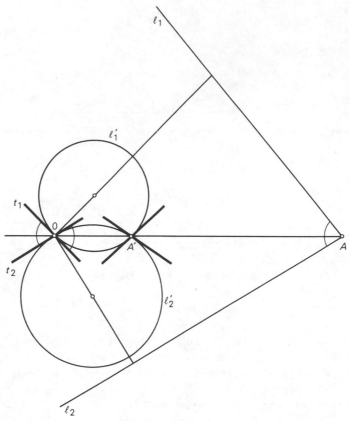

Fig. 1.27

and l_2, since the tangents t_1 and t_2 to these circles at the point O are parallel to the lines l_1 and l_2 respectively. This completes the proof of the theorem.

We leave the proofs of the following theorems to the reader as useful exercises:

THEOREM 1.7. *The angle between two circles is equal to the angle between the images of these circles under an inversion.*

THEOREM 1.8. *The angle between a circle and a straight line is equal to the angle between the images of these figures under an inversion.*

1.5. The Power of a Point with Respect to a Circle: The Radical Axis of Two Circles

The concept of the *power* of a point with reference to a circle, which is analogous to the concept of the distance from a point to a straight line, will be essential in the discussion below.

Let K be a circle of radius r in the plane, M an arbitrary point in the plane, and d the distance from M to the center O of the circle K. The *power of the point M with respect to the circle K* is defined as the number

$$S = d^2 - r^2 .$$

If the point M lies inside the circle K, then $d < r$, and the power $S = d^2 - r^2$ of M is negative. The segments of the diameter PQ on which M lies are of length $r + d$ and $r - d$. Thus, by a theorem of elementary geometry, for any chord AB of the circle K which contains M (fig. 1.28a), we have

$$S = d^2 - r^2 = -(r^2 - d^2) = -(r + d)(r - d) = -AM \cdot MB .$$

If the point M lies on the circle K, then $d = r$, and the power of M is zero. Finally, if the point M lies outside the circle K, then $d > r$ and $S = d^2 - r^2$, which is the square of the length of the tangent segment from the point M to the circle K (fig. 1.28b), is positive.

Suppose we are given two circles K_1 and K_2. The locus of points whose powers with respect to the two circles are equal is called the *radical axis* of the circles K_1 and K_2.

We have the following theorem:

THEOREM 1.9. *If K_1 and K_2 are nonconcentric circles, then their radical axis is a straight line perpendicular to the line determined by their centers.*

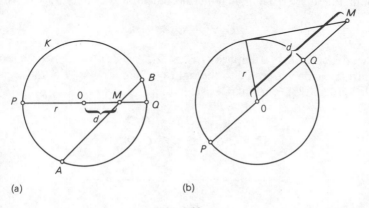

(a)   (b)

Fig. 1.28

Proof. Let O_1 and r_1, O_2 and r_2 be the centers and radii of the circles K_1 and K_2 respectively. Let d_1 and d_2 be the distances from an arbitrary point M to the centers O_1 and O_2 respectively. Then the power of M with respect to K_1 is

$$S_1 = d_1{}^2 - r_1{}^2 ,$$

and the power of M with respect to K_2 is

$$S_2 = d_2{}^2 - r_2{}^2 .$$

M lies on the radical axis of K_1 and K_2 if and only if

$$S_1 = S_2 ;$$

that is,

$$d_1{}^2 - r_1{}^2 = d_2{}^2 - r_2{}^2 .$$

This is true if and only if

$$d_1{}^2 - d_2{}^2 = r_1{}^2 - r_2{}^2 .$$

The right side of the above equation is a constant, since r_1 and r_2 are fixed. Thus the locus of the radical axis of K_1 and K_2 is the set of points M for which

$$d_1{}^2 - d_2{}^2 = k ,$$

where k is some constant and d_1 and d_2 are as defined above. Without loss of generality, we may assume that $k \geq 0$, since otherwise we can simply change the roles of the circles K_1 and K_2 and arrive at $k \geq 0$. We claim that there is a unique point S on the line of centers O_1O_2 satisfying

$$O_1S^2 - O_2S^2 = k .$$

Clearly, since $k \geq 0$ implies $O_1S \geq O_2S$, such a point S (if it exists) must coincide with or lie to the right of the midpoint H of the segment O_1O_2 (fig. 1.29). Thus, if S exists, either

(1) $O_1S + O_2S = O_1O_2$;   or

(2) $O_1S - O_2S = O_1O_2$.

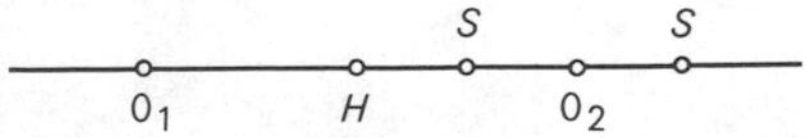

Fig. 1.29

If $0 \leq k \leq (O_1O_2)^2$, then since $k = (O_1S + O_2S)(O_1S - O_2S)$, case (1) must hold, so that there exists a unique point S on the segment HO_2 satisfying

$$k = (O_1S + O_2S)(O_1S - O_2S) = (O_1O_2)(O_1S - O_2S).$$

Analogously, if $k > (O_1O_2)^2$, case (2) holds, and there is a unique point S lying to the right of O_2 satisfying

$$k = (O_1S + O_2S)(O_1S - O_2S) = (O_1S + O_2S)(O_1O_2).$$

Now let X be an arbitrary point on the radical axis of K_1 and K_2; that is, a point of the plane satisfying

$$O_1X^2 - O_2X^2 = k.$$

Let Y be the projection of X on the line O_1O_2. By the Pythagorean theorem, we have (fig. 1.30):

$$O_1X^2 - O_1Y^2 = XY^2;$$
$$O_2X^2 - O_2Y^2 = XY^2.$$

It follows that

$$O_1X^2 - O_1Y^2 = O_2X^2 - O_2Y^2.$$

Therefore,

$$O_1Y^2 - O_2Y^2 = O_1X^2 - O_2X^2 = k. \qquad (1.1)$$

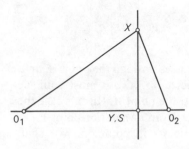

Fig. 1.30

Since Y lies on the line O_1O_2 and satisfies the relation (1.1), it must coincide with the point S. Thus, the point X lies on the perpendicular l to the line of centers O_1O_2 at S. Conversely, it is easy to show by a similar argument that all points Z on the line l satisfy

$$O_1Z^2 - O_2Z^2$$
$$= O_1Y^2 - O_2Y^2 = k.$$

The desired locus of points is thus a line perpendicular to the line of centers, and the theorem is proved.

We now consider the construction (by straightedge and compass) of the radical axis of two nonconcentric circles. As above, we assume that the circle with the larger radius is K_1, so that

$$k = r_1{}^2 - r_2{}^2 \geq 0 .$$

As shown above, if O_1 and O_2 are the centers of K_1 and K_2, and H is the midpoint of O_1O_2, then the radical axis of the circles K_1 and K_2 is perpendicular to the line O_1O_2 at the point S, which lies to the right of H. The construction of the radical axis is thus reduced to the construction of the point S on the line O_1O_2.

We shall now consider the construction of the radical axis l given the circles K_1 and K_2 in three cases:

1. K_1 and K_2 intersect at two points A and B (fig. 1.31). Since the powers of points A and B with respect to both circles must be zero, the radical axis l must coincide with the straight line AB. (In this case, the radical axis intersects the line of centers in an interior point of the segment O_1O_2.)

2. K_1 and K_2 have a unique common point A, at which they are tangent (fig. 1.32). The power of the point A with respect to the circles

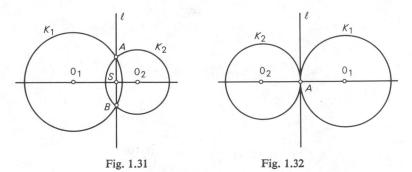

Fig. 1.31   Fig. 1.32

K_1 and K_2 is zero; thus the radical axis l passes through the point A, and, since l is perpendicular to the line of centers O_1O_2, it must coincide with the common tangent of K_1 and K_2 at the point A. (In this case the radical axis also intersects the segment O_1O_2 in an interior point.)

3. The circles K_1 and K_2 have no points in common. We shall separate this case into two subcases:

a. The circles K_1 and K_2 are situated outside one another (fig. 1.33). We draw two common tangents to K_1 and K_2, PQ and RT, with midpoints H_1 and H_2 respectively. Since the power of a point X which lies outside the circle K_1 (with respect to K_1) is equal to the square of the length of the tangent extended from X, the midpoints H_1 and H_2 have equal powers with respect to each of the circles K_1 and K_2 and, consequently, lie on and determine the radical axis l. It is easy to see that K_1 and K_2 lie on different sides of the radical axis l. (In this case, too, l intersects the line of centers at a point in the interior of the segment O_1O_2.)

b. The circle K_2 lies inside the circle K_1 (fig. 1.34). In this case, $r_1 - r_2 \geq O_1O_2$, so that

$$k = r_1{}^2 - r_2{}^2 = (r_1 + r_2)(r_1 - r_2) > (O_1O_2)^2 .$$

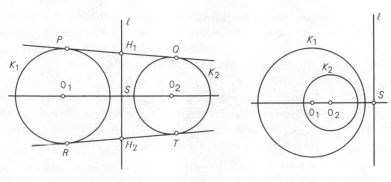

Fig. 1.33   Fig. 1.34

Thus the point S (at which the radical axis intersects the line of centers) lies to the right of O_2 (as shown above), and satisfies

$$(O_1S + O_2S)(O_1O_2) = k .$$

Setting $O_1O_2 = c$, this becomes

$$O_1S + O_2S = k/c .$$

Since S lies to the right of O_2, $O_1S = O_1O_2 + O_2S = c + O_2S$, and the equation above becomes

$$c + 2O_2S = k/c ,$$

or

$$O_2S = k/2c - c/2 .$$

Since k is constructible from r_1 and r_2 and c is given, the length O_2S is constructible; since the line O_1O_2 and the point O_2 are fixed, it follows that the point S (and thus the line l) is constructible.

In this case, the radical axis l lies outside the circle K_1, and thus both K_1 and K_2 lie to one side of l.

In each case, then, the radical axis l can be constructed by straightedge and compass from the circles K_1 and K_2.

In closing, we remark that the locus of points whose tangents to K_1 and K_2 are equal is, in cases 2 and 3, the entire radical axis, and, in case 1, all the points of the radical axis outside of the line segment AB (where A and B are the points of intersection of the circles K_1 and K_2).

1.6. Application of Inversions to the Solution of Construction Problems

The use of inversion transformations makes possible a number of elegant solutions to classical construction problems in geometry. We shall consider below problems which require the construction of a circle tangent or orthogonal to one or several circles.

I. *Problems on tangents to circles:*

Problem 1. Three nontangent circles, K_1, K_2, and K_3, intersect at some point O. We wish to construct all circles tangent to the circles K_1, K_2, K_3. It is not hard to see (fig. 1.35) that the problem has four solutions (in fig. 1.35 they are shown by dotted lines).

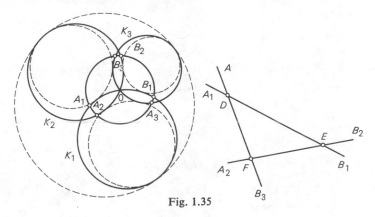

Fig. 1.35

The method of inversions allows us to find these solutions easily. Let T be an inversion with center O and radius r such that the circle of inversion intersects the circles K_1, K_2, K_3 in pairs of points A_1, B_1; A_2, B_2; A_3, B_3, respectively. Since the circles K_1, K_2, K_3 all intersect at the point O, the inversion T takes these circles into the straight lines A_1B_1, A_2B_2, and A_3B_3; since no two circles are tangent, these lines intersect pairwise. Thus our problem is reduced to constructing all circles which are tangent to the lines A_1B_1, A_2B_2, A_3B_3. Clearly, there will be one such inscribed and three such circumscribed circles for the triangle *DEF* which is formed by these lines. The construction of these circles is not difficult, and, by the rule given in sec. 1.4, we may construct the images of these four circles under the inversion *T*. This yields the required circles.

Problem 2. Construct all circles which are tangent to two given circles K_1 and K_2 and pass through a given point *O*, lying outside K_1 and K_2.

Suppose *R* is one of the desired circles. Let *T* be an inversion with center *O*. Then *T* carries K_1 and K_2 into circles K_1' and K_2' respectively, and the circle *R* into a common tangent *R'*. It is now obvious that the solutions to the problem are circles which are the images of the common tangents of the circles K_1' and K_2' under the inversion *T*. Since there are four of these tangents, the problem has four (constructible) solutions (fig. 1.36).

Problem 3 (Apollonius's problem). Construct all circles tangent to three given circles K_1, K_2, and K_3.

We shall present two solutions to this problem.

First solution. Suppose the circle *L*, with radius *R*, is one of the desired circles (fig. 1.37). We connect the segment O_1O_3 from the centers of the circles K_1 and K_3 and draw circles of radii $r_1 + s, r_2 + s$, and $r_3 + s$ around the points O_1, O_2, and O_3, respectively, where r_1, r_2, and r_3 are the radii of K_1, K_2, K_3, and

$$ s = \frac{(O_1O_3 - r_1 - r_3)}{2} . $$

We denote the constructed circles by $\bar{K}_1$, $\bar{K}_2$, $\bar{K}_3$, respectively. Let $\bar{L}$ be the circle concentric with *L* having radius $\bar{R} = R - s$. It is clear that if we can construct the circle $\bar{L}$, we can easily construct the circle *L*. It is obvious that $\bar{L}$ is tangent to the circles $\bar{K}_1$, $\bar{K}_2$, $\bar{K}_3$. The circles $\bar{K}_1$ and $\bar{K}_3$ are constructed so that they are tangent to one another at some point *D*. Let *T* be an inversion with center *D* and radius *r* such that the circle of inversion intersects the circles $\bar{K}_1$ and $\bar{K}_3$. The inversion *T* takes the circles $\bar{K}_1$ and $\bar{K}_3$ into a pair of parallel lines l_1 and l_3, and the circle $\bar{K}_2$ into a circle $\bar{K}_2'$. The circle $\bar{L}$ is taken by the inversion *T* into a circle $\bar{L}'$, which is tangent to $\bar{K}_2'$ and to both the parallel lines l_1 and l_3. In this way, the solution of Apollonius's problem has been reduced to a simpler construction problem: to construct all circles tangent to a given pair of parallel lines and to a given circle.

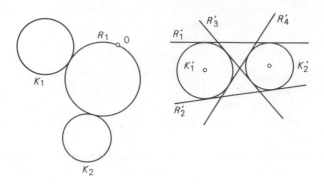

Fig. 1.36

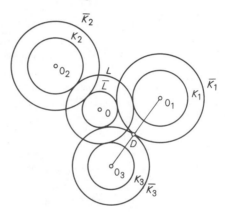

Fig. 1.37

We leave the solution of this problem to the reader and suggest that the reader verify that the pair of circles K_1 and K_2 or K_2 and K_3 could be used in place of the pair K_1 and K_3 in the above construction.

Second solution. We shall perform an auxiliary construction that will reduce Apollonius's problem to problem 2. Suppose, without loss of generality, that the circle K_3 has radius r_3 satisfying $r_1 \geq r_3$ and $r_2 \geq r_3$. Suppose L is one of the circles tangent to the circles K_1, K_2, and K_3. We construct the circles $\bar{K}_1$ and $\bar{K}_2$, with centers O_1 and O_2 and radii $\rho_1 = r_1 - r_3$ and $\rho_2 = r_2 - r_3$, respectively (fig. 1.38). The circle $\bar{L}$, constructed with center O and radius $\rho = R + r_3$, where R is the radius of L, will be tangent to $\bar{K}_1$ and $\bar{K}_2$ and will pass through the point O_3. Construction of the circle $\bar{L}$ is given in the solution of problem 2.

The constructed circle $\bar{L}$ is concentric with the desired circle L and has a radius which is larger by r_3. The rest of the solution is left to the reader as an exercise.

II. *Construction of a circle which intersects given circles orthogonally:* We shall say that two curves intersect *orthogonally* at a point M or that they are *orthogonal* at the point M if the tangents to these curves at the point M are perpendicular.

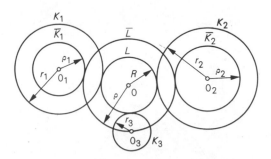

Fig. 1.38

Problem 4. Given two nonconcentric circles K_1 and K_2, we wish to construct all circles orthogonal to K_1 and K_2 passing through a given point M.

The solution to this problem is broken down into a number of cases depending on the relative position of the circles K_1, K_2, and the point M:

a. The circles K_1 and K_2 intersect at two points A and B (fig. 1.39a). It is obvious that if M coincides with one of the points A or B, then the desired circle k can exist only if one of the circles we are considering has zero radius. Therefore, in what follows we shall examine the case where the point M is distinct from the points A and B.

Let T be an inversion transformation with center A and radius $r = AB$. Then T takes the point M into some point M', the point B remains unchanged, and the circles K_1 and K_2 are transformed into distinct straight lines K_1 and K_2 passing through the point B (fig. 1.39b). The image k' of the desired circle k under T must be a circle or a straight line orthogonal to the nonparallel lines K_1' and K_2' and passing through the point M', which is distinct from A and B. It is obvious that there is only one circle satisfying these conditions (there is no line k' satisfying the above conditions). This circle has center B and radius BM'. We denote this circle by k' (fig. 1.39b). Since two iterations of the inversion T yield the identity transformation, the image of the circle k' under T is the desired circle k. In solving the problem in this case we

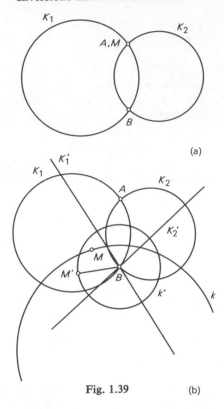

(a)

(b)

Fig. 1.39

have established that there is a unique solution regardless of the position of the point M.

b. The circles K_1 and K_2 are tangent at a single point A.

If the point M coincides with the point A, the problem has infinitely many solutions: first, the line of centers O_1O_2 of the circles K_1 and K_2 (fig. 1.40), and, second, any circle with its center on the common tangent of K_1 and K_2 which passes through the point A.

Now let M be any point in the plane other than A. Let T denote the inversion transformation with center A and radius $r = AM$. Then the inversion T fixes the point M and takes the circles K_1 and K_2 into the parallel lines K'_1 and K'_2 (fig. 1.41). The image k' of the desired circle k under T should be either a circle or a straight line, passing through the point M and orthogonal to the parallel lines K'_1 and K'_2. Clearly, k' must be a line (and not a circle). Since the line k' must pass through the fixed point M and must be perpendicular to the two parallel lines K'_1 and K'_2, it is uniquely determined. Inversion by T takes the line k' into the desired circle k.

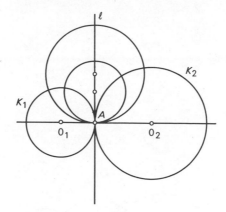

Fig. 1.40

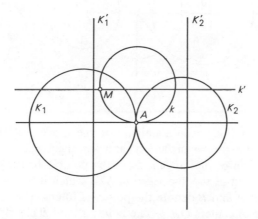

Fig. 1.41

Thus, in this case, if the point M is different from the point A, the problem has a unique solution.

c. The circles K_1 and K_2 have no points in common. We claim that there is a point A on the line of centers O_1O_2 and an inversion T with center A (fig. 1.42) which transforms the circles K_1 and K_2 into a pair of concentric circles.

Let l be the radical axis of the circles K_1 and K_2. Let S be the point of intersection of l with the line of centers O_1O_2. As we showed in sec. 1.5, since K_1 and K_2 have no points in common, the point S lies outside both circles K_1 and K_2. We draw a tangent from S to the circle K_1, with point of tangency T_1. The circle K with center S and radius $R = ST_1$ intersects

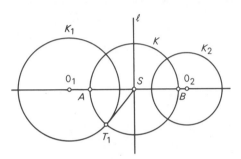

Fig. 1.42

the circles K_1 and K_2 orthogonally. For the circles K_1 this follows immediately from the construction, and for the circle K_2 it follows because the length of the tangent from the point S to the circle K_2 is equal to the length of the line segment ST_1, which is the radius of the circle K. We let A and B denote the points of intersection of the circle K with the line of centers O_1O_2. The points A and B clearly do not lie on either of the circles K_1 or K_2.

The inversion T takes the following form: We place the center of T at the point A, and we take the radius r to be equal to the length of the line segment AB; that is, $r = AB$.

The inversion T leaves the point B fixed; takes the circle K into the line K', which passes through the point B and is perpendicular to the line of centers O_1O_2; leaves the line of centers O_1O_2 invariant; and takes the circles K_1 and K_2 into circles K_1' and K_2', whose centers lie on the line O_1O_2 (fig. 1.43). Since the line K' is orthogonal to both circles K_1' and K_2', the centers of K_1' and K_2' must lie on K'. It follows that the centers of the circles K_1' and K_2' lie on the point of intersection of the lines K' and O_1O_2—that is, that K_1' and K_2' are concentric circles with **center B.**

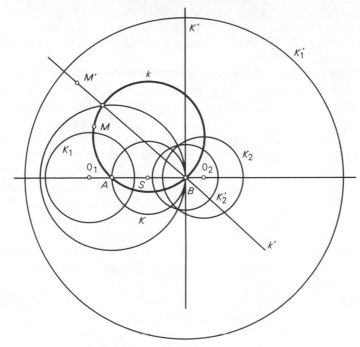

Fig. 1.43

Now we assume that the point M is distinct from the points A and B. Then M', its image under the inversion T, is also distinct from these points. If k' is the image under T of one of the desired circles k, then k' must be a line passing through the points B and M'. It follows that the line k' is unique. Applying the inversion T, we obtain the desired circle k. Thus, if the point M is distinct from the points A and B, the problem has a unique solution. If M coincides with the point B, then we can take any line passing through B for k'. In this instance, then, the problem has infinitely many solutions.

If the point M coincides with the point A, the problem again has an infinite number of solutions. To show this, it is sufficient to do the above constructions with one substitution: we consider the inversion T_1 with center at B and radius $r = AB$.

In this way, we have considered all the possible relative positions of the point M and the circles K_1 and K_2. The problem has been solved completely.

Problem 5. Given three circles K_1, K_2, K_3, situated so that each lies outside the other two, construct all circles which are orthogonal to all three given circles.

Solution. By the assumptions, the circles K_1, K_2, K_3 are situated so that the radical axis of any pair of them separates the corresponding circles. Therefore, the pairs K_1 and K_2, K_2 and K_3 have radical axes l_1 and l_2 which are not coincident.

There are two possible cases:

a. The lines l_1 and l_2 are parallel. Then the centers of the circles K_1, K_2, K_3 are collinear. The line on which they lie is the solution to the problem.

b. The lines l_1 and l_2 intersect at some point S. By assumption, the circles K_1, K_2, K_3 are situated so that their radical axes lie outside the corresponding pairs of circles. Therefore, we can draw tangents from the point S to each of the circles K_1, K_2, K_3. All the tangents have equal lengths. Let ST_1 be a tangent to the circle K_1 (where T_1 is the point of tangency) and let r be the length of the tangent. The circle with center at S and radius r is clearly the circle we are seeking.

From these considerations, it follows that the problem always has only one solution. We leave it to the reader to verify this fact.

1.7. Pencils of Circles

If K_1 and K_2 are two circles in the plane, the set of all circles orthogonal to K_1 and K_2 is called the *pencil of circles produced by K_1 and K_2* and is denoted by $P(K_1, K_2)$. Often, if the circles K_1 and K_2 do not play an important role in the pencil produced, we denote the pencil simply by P or Q. Since we decided above to consider straight lines as special cases of circles, straight lines, as well as circles, can enter into the production of pencils.

We now consider three pencils arranged in the simplest ways. These pencils arise from special arrangements of the circles K_1 and K_2:

1. K_1 and K_2 are concentric circles with common center B. In this case, the pencil $P(K_1, K_2)$ is clearly the set of all straight lines passing through the point B (fig. 1.44). This pencil is called an *elementary elliptical pencil.*

2. K_1 and K_2 are straight lines intersecting at the point B. The pencil $P(K_1, K_2)$ is clearly the set of all concentric circles with common center B (fig. 1.45). This pencil is called an *elementary hyperbolic pencil.*

3. K_1 and K_2 are parallel lines. The pencil $P(K_1, K_2)$ clearly consists of all the lines perpendicular to the lines K_1 and K_2 (fig. 1.46). This pencil is called an *elementary parabolic pencil.*

We now consider how the various elementary pencils differ from one another.

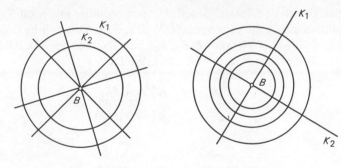

Fig. 1.44   Fig. 1.45

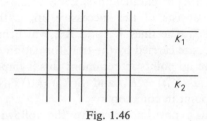

Fig. 1.46

Type of pencil	Number of common points in the circles K_1 and K_2
Elliptical	0
Parabolic	1 (the point at infinity)
Hyperbolic	2 (the point B and the point at infinity)

Since circles (including lines) can have no more than two points in common, it is clear that there are in some sense only three different "types" of elementary pencils.

More precisely, we shall show that for any pair of circles K_1 and K_2 we can transform the pencil $P(K_1, K_2)$ into one of the three elementary pencils by application of a properly chosen inversion. Furthermore, since inversions are one-to-one transformations, any pencil P can be transformed by an inversion into an elementary pencil of only one definite type. For example, if the inversion T takes the pencil $P(K_1, K_2)$

into the elementary elliptical pencil P', then no other inversion T_1 can take it into a parabolic or a hyperbolic pencil P_1. This can be demonstrated as follows: If T_1 takes $P(K_1, K_2)$ into P_1, then, on the basis of theorem 1.1, T_1 takes P_1 into $P(K_1, K_2)$. We let

$$K_1' = T(K_1), \qquad K_1'' = T_1(K_1),$$
$$K_2' = T(K_2), \qquad K_2'' = T_1(K_2).$$

Then

$$K_1 = T_1(K_1''), \qquad K_2 = T_1(K_2'').$$

Since P' is an elementary elliptical pencil, and P_1 is an elementary parabolic or hyperbolic pencil, K_1' and K_2' are concentric circles, and K_1'' and K_2'' are intersecting or parallel lines. Let S be the transformation of the plane which consists of the successive applications of the two inversions T_1 and T. The lines K_1'' and K_2'', which have at least one common point O_∞, are carried by the transformation S into the circles K_1', K_2', which have no points in common; this is impossible, since the figures $S(K_1'') = T(T_1(K_1'')) = K_1'$ and $S(K_2'') = T(T_1(K_2'')) = K_2'$ must have at least one point in common.

We are now in a position to prove the following fundamental theorem.

THEOREM 1.10. a. *If the circles K_1 and K_2 have no points in common, then there exists an inversion or identity transformation T_1 carrying $P(K_1, K_2)$ into an elementary elliptical pencil.*

b. *If the circles K_1 and K_2 have a unique common point, then there exists an inversion or identity transformation T_2 carrying $P(K_1, K_2)$ into an elementary parabolic pencil.*

c. *If the circles K_1 and K_2 have two common points, then there exists an inversion or identity transformation T_3 carrying $P(K_1, K_2)$ into an elementary hyperbolic pencil.*

The proof of theorem 1.10 is closely related to the constructions we performed in sec. 1.6 in the solution to problem 4. Subsequent constructions will depend on the following lemma.

LEMMA 1.2. *Suppose the inversion T carries the circles K_1 and K_2 into the circles K_1' and K_2' respectively. Then the image of the pencil $P(K_1, K_2)$ under T is the pencil $P(K_1', K_2')$.*

Proof of the lemma. Since any inversion preserves the orthogonality of circles, the image of $P(K_1, K_2)$ under T is contained in the pencil

$P(K_1', K_2')$. Therefore, to prove that the image of $P(K_1, K_2)$ coincides with the pencil $P(K_1', K_2')$, it is sufficient to show that the pencil $P(K_1', K_2')$ is contained in the image of $P(K_1, K_2)$; that is, that for any circle k' of the pencil $P(K_1', K_2')$, there is a circle k in the pencil $P(K_1, K_2)$ such that $T(k) = k'$. If k' is a circle in $P(K_1', K_2')$, let

$$k = T(k') .$$

The circle k is orthogonal to K_1 and K_2 and thus lies in the pencil $P(K_1, K_2)$. Since an inversion executed twice in succession is the identity transformation, we obtain

$$T(k) = T(T(k')) = k' ,$$

and the lemma is proved.

1. *Proof of statement a.* Let K_1 and K_2 be two circles having no points in common. If K_1 and K_2 are concentric, $P(K_1, K_2)$ is an elementary elliptical pencil, and we may choose T_1 as the identity transformation. The interesting case is that in which the circles are not concentric (fig. 1.47). One of the circles, K_1 or K_2, may be a straight line (but not both, since then K_1 and K_2 would have at least one point in common, the point O_∞).

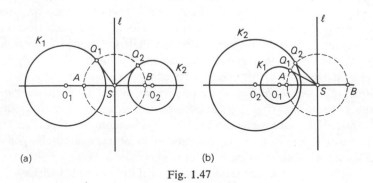

(a)   (b)

Fig. 1.47

First, suppose that neither K_1 nor K_2 is a straight line. Let S be the intersection of the radical axis l with the line of centers O_1O_2 of the circles K_1 and K_2 (constructing the point S and the line l as in sec. 1.5). The line l and, consequently, the point S lie outside both circles K_1 and K_2; therefore, we can draw tangents SQ_1 and SQ_2 from the point S to the circles K_1 and K_2 (with Q_1 and Q_2 the corresponding points of tangency). The point S lies on the radical axis of K_1 and K_2; therefore $SQ_1 = SQ_2$. The circle K, with center S and radius $a = SQ_1$, intersects K_1 and K_2 orthogonally. Let A and B be the points of intersection of K with the line of centers O_1O_2.

We define the inversion T_1 to have center A and radius AB. In problem 4 of sec. 1.6, it was proved that the inversion T_1 transforms the circles K_1 and K_2 into concentric circles K_1' and K_2' with common center B. By the lemma, the pencil of circles $P(K_1, K_2)$ is taken by the inversion T_1 into the pencil $P(K_1', K_2')$, which consists of all lines passing through the point B.

Thus, the inversion T_1 transforms the pencil $P(K_1, K_2)$ into an elementary elliptical pencil.

It remains to consider the case when one of the circles, say K_1, is a straight line (fig. 1.48). Since K_1 and K_2 have no points in common, K_1 lies outside K_2. We construct a line m through O_2 perpendicular to K_1, and let S be the point at which m intersects K_1. We construct further a tangent SQ_2 to K_2. Let K be the circle with center S and radius $a = SQ_2$, and A and B the points of intersection of K with the line m. The inversion T_1 having center A and radius $r = AB$ leaves the point B fixed, leaves the line m invariant, and takes the circle K into the line K', which passes through the point B and is perpendicular to the line m.

The line K_1 does not pass through the point A, and the circle K is orthogonal to the line K_1 and the circle K_2. Therefore, the images of K_1 and K_2 under T_1 will be the circles K_1' and K_2', whose centers lie simultaneously on the lines K' and m; that is, K_1' and K_2' are concentric circles with center B. It follows (by the lemma) that the image of the pencil $P(K_1, K_2)$ is the elementary elliptical pencil $P(K_1', K_2')$.

Thus statement a is completely proved.

Proof of statement b. Let K_1 and K_2 be two circles having exactly one point A in common (fig. 1.49). If both K_1 and K_2 are straight lines, they must be parallel, since they can have no common points other than O_∞. In this case, then, $P(K_1, K_2)$ is already an elementary parabolic pencil, and we can choose T_2 to be the identity transformation.

If neither K_1 nor K_2 are straight lines, or if only one of them (say K_1) is a straight line, we may choose any inversion T_2 with center A. K_1 and K_2 are transformed by T_2 into parallel lines K_1' and K_2'; thus the image $P(K_1', K_2')$ of $P(K_1, K_2)$ under T_2 is an elementary parabolic pencil.

This completes the proof of statement b.

Proof of statement c. Let K_1 and K_2 be two circles having two points A and B in common (fig. 1.50). If both K_1 and K_2 are straight lines, then they have exactly one point of intersection other than the point O_∞, and $P(K_1, K_2)$ is already an elementary hyperbolic pencil (so that we may choose T_3 to be the identity transformation).

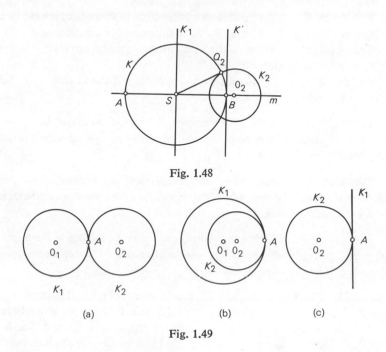

Fig. 1.48

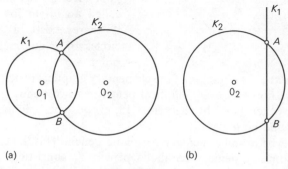

(a)   (b)   (c)

Fig. 1.49

If at least one of K_1 and K_2 is not a straight line, let T_3 be the inversion with center A and radius $r = AB$. Then the images of K_1 and K_2 under T_3 will be the lines K_1' and K_2' intersecting at the point B (fig. 1.51). It

(a)   (b)

Fig. 1.50

follows that the image $P(K_1', K_2')$ of $P(K_1, K_2)$ under T_3 is an elementary hyperbolic pencil.

Thus statement c is proved, and the proof of theorem 1.10 is complete.

We now introduce the following definitions:

The pencil $P(K_1, K_2)$ produced by the circles K_1 and K_2 is said to be *elliptical* if the circles K_1 and K_2 have no points in common.

The pencil $P(K_1, K_2)$ is said to be *parabolic* if the circles K_1 and K_2 have exactly one common point.

The pencil $P(K_1, K_2)$ is said to be *hyperbolic* if the circles K_1 and K_2 have two common points.

THEOREM 1.11. *Every elliptical pencil can be obtained from some elementary elliptical pencil by the application of an appropriate inversion or identity transformation.*

THEOREM 1.12. *Every parabolic pencil can be obtained from some elementary parabolic pencil by the application of an appropriate inversion or identity transformation.*

THEOREM 1.13. *Every hyperbolic pencil can be obtained from some elementary hyperbolic pencil by the application of an appropriate inversion or identity transformation.*

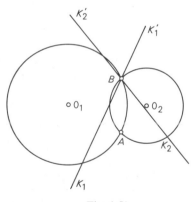

Fig. 1.51

The proofs of theorems 1.11, 1.12, and 1.13 follow immediately from theorem 1.10 and the fact that two successive applications of the same inversion yield the identity transformation on the plane.

The point A is called a *node* for the pencil P if all the circles of P pass through A. The point A is called an *origin* for the pencil P if there exists a sequence of circles of P contracting into the point A.

From the construction of the elementary elliptical pencil and theorem 1.11, we find that every elliptical pencil has two nodes and no origin. On the other hand, by theorem 1.13, every hyperbolic pencil has two origins and no nodes.

Let P be a nonelementary parabolic pencil. This pencil is obtained from some elementary parabolic pencil P', consisting of a class of parallel lines, under transformation by an inversion T. Let A be the center of the inversion *T*. It is not hard to see that *P* is the set of all circles mutually tangent at the point *A*, including the common tangent of all the circles at the point *A* (fig. 1.52). Thus, the pencil *P* has one node and one origin, both of which are the point *A*. Operating on the pencil *P* by the inversion *T*, we obtain the elementary parabolic pencil *P'*, for which the point O_∞ is both the only node and the only origin.

From the above discussion, we obtain:

THEOREM 1.14. *The total number of nodes and origins for any pencil is two.*

The pencil *P* is said to be *orthogonal* to the pencil *Q* if any circle in the pencil *P* is orthogonal to any circle in the pencil *Q*. It is obvious that, if the pencil *P* is orthogonal to the pencil *Q*, then, conversely, the pencil *Q* is orthogonal to the pencil *P*.

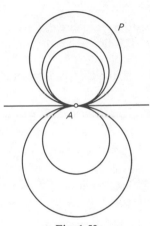

Fig. 1.52

We now consider pairs of orthogonal elementary pencils. If *P* is an elementary elliptical pencil, that is, the set of all lines passing through some point *B*, then the set of all circles orthogonal to the circles of *P* is clearly an elementary hyperbolic pencil *Q*, consisting of all concentric circles with center *B* (we adjoin to *Q* the point *B* and the point O_∞, which are the origins of *Q*). It is easy to see that, conversely, the pencil *Q* is orthogonal to *P*, and in addition, that the nodes of *P* are the origins of *Q*.

If *P* is an elementary parabolic pencil—a set of parallel lines together with the point O_∞—then the pencil *Q*, obtained by rotating *P* through a right angle, will be orthogonal to *P*, and conversely. Thus the nodes and origins of the pencils *P* and *Q* coincide at O_∞.

From the above discussion and from theorems 1.11, 1.12, and 1.13, we obtain the following theorem:

THEOREM 1.15. *For every pencil P there exists one and only one orthogonal pencil Q. If P is an elliptical pencil, then Q is a hyperbolic pencil, and conversely: the nodes of P are the origins of Q, and conversely. If P is a parabolic pencil, then Q is also a parabolic pencil. In this case, the nodes and origins of the pencils P and Q coincide at a single point A. The pencil Q is obtained from the pencil P by rotating the pencil P through a right angle about the point A.*

1.8. Structure of an Elliptical Pencil

THEOREM 1.16. *Every elliptical pencil P is the set of all circles passing through some two fixed points.*

Proof. If P is an elementary elliptical pencil with node B, then P is the set of all circles passing through the points B and O_∞. If P is non-elementary, there exists an elementary elliptical pencil P' and an inversion T (see theorem 1.11), carrying the pencil P' into the pencil P.

P' is the set of straight lines passing through some point B' (fig. 1.53). Let A be the center of the inversion T. Then A and B' are distinct; if not, the inversion T would carry the pencil P' into itself, and P would be elementary. Since the image of the pencil P' under the inversion T is the set of circles passing through the points A and $B = T(B')$, the theorem is proved.

COROLLARY 1. *The points A and B are nodes of the pencil P.*

Thus, every elliptical pencil can be defined as the set of circles passing through two fixed points (nodes of the pencil). It follows that the nodes uniquely define the elliptical pencil.

If one of the given nodes is the point at infinity, the elliptical pencil is elementary.

COROLLARY 2. *Let A and B be the nodes of the pencil P. Then the straight line AB is an element of the pencil P.*

If A and B are ordinary points, then the line AB is the only straight line in the pencil P (all the other elements of P are circles). It is easy to see that the line AB is the radical axis for any pair of circles in the pencil P. Therefore, the line AB is called the *radical axis of the pencil P.*

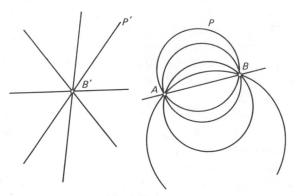

Fig. 1.53

Thus, a nonelementary elliptical pencil is the set of all "real" circles passing through two fixed points, and the common radical axis of all pairs of circles taken from this set. As noted, this radical axis passes through the nodes of the elliptical pencil.

If one of the points A and B, say A, is the point at infinity, then the pencil P consists of all straight lines passing through the point B. In this case, the uniqueness of the line AB disappears, and, consequently, for an elementary elliptical pencil the concept of a radical axis becomes meaningless. Thus, the presence of exactly one straight line in an elliptical pencil is a necessary and sufficient condition for the pencil to be nonelementary.

1.9. Structure of a Parabolic Pencil

THEOREM 1.17. *Every nonelementary parabolic pencil P is the set of all circles tangent to one another at some fixed point.*

Proof. Since P is a nonelementary parabolic pencil, there exist an elementary parabolic pencil P' and an inversion T (see theorem 1.12) carrying P' to P. P' is a class of mutually parallel lines to which is added the point at infinity. Let A be the center of the inversion T, and l the straight line in P' passing through the point A. Then the inversion T leaves l invariant and transforms all

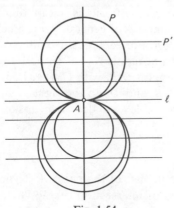

other lines of the pencil P' into circles tangent to l at the point A (fig. 1.54). Since the image under T of the point O_∞ is A, it follows that the pencil P is the set of all circles tangent to one another at the point A, and that the point A is the origin of the pencil P. This proves the theorem.

We note that if P is an elementary parabolic pencil, that is, a class of parallel lines, then P is a set of circles tangent at the point O_∞.

Fig. 1.54

COROLLARY. *The straight line l is an element of the pencil P.*

The line l is the radical axis of any pair of circles of the pencil P. Therefore, l is called the *radical axis of the pencil P.*

It is clear from theorem 1.17 that every nonelementary parabolic pencil can be defined by its node (or, since they are the same, its origin) A and the radical axis l passing through that point.

If the node of the parabolic pencil is the point at infinity, then it is an elementary parabolic pencil, and isolation of a radical axis is meaningless.

Just as in the case of elliptical pencils, a necessary and sufficient condition that a parabolic pencil be nonelementary is that it contain a unique straight line, the radical axis of the pencil.

1.10. Structure of a Hyperbolic Pencil

Hyperbolic pencils have a more complicated structure than the elliptical and parabolic pencils described in secs. 1.8 and 1.9.

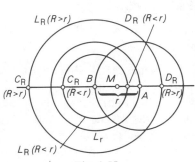

Fig. 1.55

Let P be an arbitrary nonelementary hyperbolic pencil. From theorem 1.13 it follows that there exists an elementary hyperbolic pencil P' and an inversion T which carries P' to P. The pencil P' is the set of all concentric circles with a common center at some point B (fig. 1.55). Let A be the center of the inversion T and r its radius. From the proof of theorem 1.10, it is clear that, without loss of generality, r can be chosen as the length of the line segment AB. For each positive number R, let L_R denote the circle with center B and radius R. Let C_R and D_R be the points of intersection of L_R with the line AB, with C_R (fig. 1.55) regarded as lying to the left of the point B, and D_R to the right of B. Let K_R (fig. 1.56) denote the image of the circle L_R under the inversion T. We first consider the case where

$$R < r;$$

in this case both points C_R and D_R lie to the left of the point A. Their images C'_R and D'_R, which are the points of intersection of the circle K_R with the line AB, also lie to the left of the point A. Furthermore,

$$AC_R = r + R > r = AB > AD_R = r - R,$$

and, therefore,

$$\frac{r}{2} < \frac{r^2}{r + R} = AC'_R < AB = r < \frac{r^2}{AD_R} = \frac{r^2}{r - R} = AD'_R.$$

It follows that the point C'_R lies in the interior of the segment BM, where M is the midpoint of the segment AB; that the point D'_R lies outside the line segment AB to the left of the point B; and, finally, that the center of the circle K_R is located at the point Q_R, also lying to the left of B, since

$$AQ_R = \frac{C'_R A + D'_R A}{2} = \frac{r^2}{2}\left(\frac{1}{r+R} + \frac{1}{r-R}\right) = \frac{r^3}{r^2 - R^2} > r \,.$$

If $R = r$, the circle $L_R = L_r$ passes through the point A (fig. 1.55), and since

$$AC'_r = \frac{r^2}{AC_r} = \frac{r^2}{2r} = \frac{r}{2}\,,$$

the inversion T takes L_r into the straight line K_r, which is perpendicular to the line segment AB at its midpoint M (fig. 1.56).

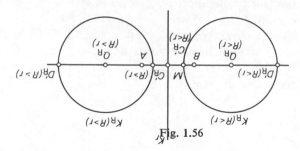

Fig. 1.56

If $R > r$, then the point C_R lies to the left of B, and the point D_R lies to the right of A (fig. 1.55). Since

$$AC'_R = \frac{r^2}{AC_R} = \frac{r^2}{r+R} < \frac{r}{2} = AM\,,$$

the point C'_R lies in the interior of the line segment AM, and the point D'_R lies outside the line segment AB to the right of A. The entire circle K_R thus lies to the right of the line K_r (fig. 1.56), and its center, the point Q_R, lies to the right of the point A, since $AC'_R < AD'_R$. This can be shown as follows:

$$AC'_R = \frac{r^2}{r+R} < \frac{r^2}{R-r} = \frac{r^2}{AD_R} = AD'_R\,.$$

Let $h(R)$ denote the radius of the circle K_R.

If $R < r$, then

$$h(R) = \frac{D'_R A - C'_R A}{2} = \frac{r^2}{2}\left(\frac{1}{r-R} - \frac{1}{r+R}\right) = \frac{r^2 R}{(r+R)(r-R)}.$$
(1.2)

As R converges to r, it follows from formula (1.2) that $h(R)$ increases without bound. A simple visual picture corresponds to this: The circles K_R of the hyperbolic pencil P expand without bound as the parameter R increases from 0 to r and, for $R = r$, become the straight line K_r.

If $R > r$, then

$$h(R) = \frac{C'_R A + D'_R A}{2} = \frac{r^2}{2}\left(\frac{1}{r+R} + \frac{1}{R-r}\right) = \frac{r^2 R}{(R-r)(R+r)}.$$
(1.3)

It follows that as R approaches r from above, the circles K_R expand without bound and, for $R = r$, become the line K_r. If R increases monotonically from r to $+\infty$, it follows from formula (1.3) that the circles K_R contract (their radii approach zero). For $R = +\infty$, the circle K_R becomes the point A.

The general form of a hyperbolic pencil P is represented in figure 1.57. We remark that the straight line K_r is the radical axis of any pair of circles in the pencil P. (We leave the proof of this fact to the reader.) The line K_r is thus called the *radical axis of the pencil P*.

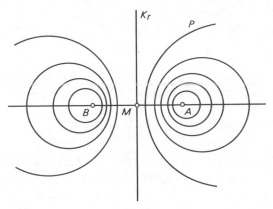

Fig. 1.57

From the above discussion it is clear that a hyperbolic pencil is completely specified by its origins or by one of its origins and its radical axis.

If one of the origins is the point at infinity, then the pencil P is an elementary hyperbolic pencil; that is, a set of concentric circles. The concept of a radical axis loses its meaning for such a pencil.

Since an elementary hyperbolic pencil contains no straight lines, a necessary and sufficient condition for a hyperbolic pencil to be non-elementary is the presence of a straight line. As we know, in a non-elementary hyperbolic pencil, this line is unique.

1.11. Ptolemy's Theorem

In this section we shall investigate the problem of determining when it is possible to pass a circle through four given points in a plane. It happens that this question can be partially answered with the help of the well-known theorem of Ptolemy from elementary geometry. We shall formulate and prove the theorem of Ptolemy a little later; first, let us consider the solution of the problem by means of inversions.

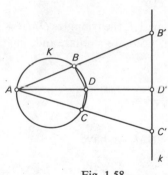

Fig. 1.58

Let A, B, and C be three noncollinear points in the plane. There is a unique circle K passing through these points (fig. 1.58). Let T be an inversion with center A and some radius r which is greater than the diameter of the circle K. The image of the circle K under the inversion T will be a line k which lies completely outside of K, since r is greater than the diameter of K. Let B' and C' denote, as usual, the images of the points B and C under T. The points B' and C' clearly lie on the line k. Now we take an arbitrary point D in the plane, and let D' be its image.[2] If the point D lies on the circle K, the point D' will lie on the line k; if D does not lie on K, then D' will not lie on k. Therefore, in order for the four points A, B, C, D to lie on the circle K, it is necessary and sufficient that the points B', C', and D' lie on the line k.

If the three distinct points B', C', and D' are collinear, then the segments $B'C'$, $C'D'$, and $B'D'$ satisfy one and only one of the three relations:

$$B'D' + D'C' = B'C';$$
$$B'C' + C'D' = B'D'; \qquad (1.4)$$
$$C'B' + B'D' = C'D'.$$

2. We assume that the point D is distinct from the points A, B, and C.

If the three points B', C', and D' are not collinear, then the inequality

$$B'D' + C'D' > B'C' \qquad (1.5)$$

holds.

We shall now attempt to write the relations (1.4) and (1.5) so that they do not involve the points B', C', and D'.

As a preliminary, we establish the following lemma:

LEMMA 1.3. *Let the inversion T with center O and radius r be given. Let M and N be two arbitrary points in the plane different from O and from the point O_∞. Then*

$$M'N' = MN \frac{r^2}{OM \cdot ON},$$

where

$$M' = T(M); \qquad N' = T(N).$$

Proof. By lemma 1.1, the triangles OMN and $ON'M'$ (fig. 1.59) are similar, and, in particular,

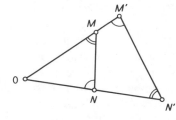

Fig. 1.59

$$\frac{M'N'}{MN} = \frac{OM'}{ON}.$$

Since $OM' = r^2/OM$, we have

$$M'N' = MN \frac{r^2}{OM \cdot ON},$$

and the lemma is proved.

From lemma 1.3, we have

$$B'D' = BD \cdot \frac{r^2}{AB \cdot AD}, \quad D'C' = DC \cdot \frac{r^2}{AD \cdot AC}, \quad B'C' = BC \cdot \frac{r^2}{AB \cdot AC}.$$

Thus, if the points A, B, C, and D lie on the circle K, the images of B, C, and D lie on the line k, and the relation

$$BD \cdot \frac{r^2}{AB \cdot AD} + DC \cdot \frac{r^2}{AD \cdot AC} = BC \cdot \frac{r^2}{AB \cdot AC}$$

is valid. (We assume, without loss of generality, that D' lies between B'

and C'.) If the points A, B, C, and D do not lie on the circle K, then the relation

$$BD \cdot \frac{r^2}{AB \cdot AD} + DC \cdot \frac{r^2}{AD \cdot AC} > BC \cdot \frac{r^2}{AB \cdot AC}$$

is valid.

It follows that

$$BD \cdot AC + DC \cdot AB = BC \cdot AD \,,$$

if the points A, B, C, D lie on one circle, and

$$BD \cdot AC + DC \cdot AB > BC \cdot AD \,,$$

if the points A, B, C, D do not lie on one circle.

Thus, we have:

THEOREM 1.18. *In order that the four points A, B, C, D lie on one circle and that the points A and D lie on different arcs with endpoints B and C, it is necessary and sufficient that the equality*

$$BD \cdot AC + DC \cdot AB = BC \cdot AD$$

be satisfied.

Since any quadrilateral $ABCD$ inscribed in a circle K satisfies the conditions of theorem 1.18, we have:

THEOREM 1.19 (Ptolemy's theorem). *For every quadrilateral inscribed in a circle, the sum of the products of the opposite sides is equal to the product of the diagonals.*

Complex Numbers and Inversions

2.1. Geometric Representation of Complex Numbers and Operations on Them

As we know, every complex number $z = x + iy$ (where i is the *imaginary* unit defined by the relation $i^2 = -1$) can be conveniently represented in the Cartesian plane by the ordered pair of coordinates (x, y). (We assume that the coordinate axes of the plane are fixed with origin O, as in fig. 2.1.) For every point M in the plane there is a unique vector **r** with initial point O and terminal point M. This vector is called the *radius vector* of the point M, and the coordinates of the point M are called the *coordinates* or *components* of the radius vector. Therefore, the complex number $z = x + iy$ can be represented geometrically by the radius vector with coordinates (x, y).

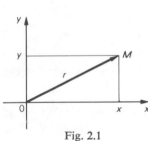

Fig. 2.1

If $z_1 = x_1 + iy_1$ and $z_2 = x_2 + iy_2$ are two complex numbers, and **r**$_1$ and **r**$_2$ are their corresponding radius vectors, then the numbers $z_1 + z_2$ and $z_1 - z_2$ are defined by:

$$z_1 + z_2 = (x_1 + x_2) + i(y_1 + y_2);$$
$$z_1 - z_2 = (x_1 - x_2) + i(y_1 - y_2).$$

On the other hand, from the definition of the rules for addition and subtraction of vectors (we have in mind the parallelogram rule), it follows that the vectors **r**$_1$ + **r**$_2$ and **r**$_1$ - **r**$_2$ have coordinates $(x_1 + x_2, y_1 + y_2)$ and $(x_1 - x_2, y_1 - y_2)$ respectively. Therefore, the addition and subtraction of two complex numbers can be performed on

48

their radius vectors by taking the corresponding sum and difference of the radius vectors representing the given complex numbers (fig. 2.2).

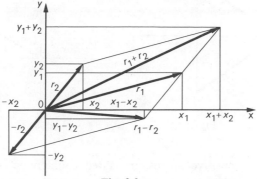

Fig. 2.2

The number $\bar{z} = x - iy$ is called the *conjugate* of the number $z = x + iy$. Let M be the endpoint of the radius vector $\mathbf{r}$ corresponding to the number $z = x + iy$, and let M_1 be the endpoint of the radius vector $\mathbf{r}_1$ corresponding to the number $\bar{z} = x - iy$. Since the points M and M_1 have as their coordinates (x, y) and $(x, -y)$ respectively, M_1 can be obtained from M by reflection across the x-axis (fig. 2.3).

Let z be some complex number and $\mathbf{r}$ its radius vector. Let $|z|$ denote the length of the vector $\mathbf{r}$, and φ the angle measured counterclockwise from the positive side of the x-axis to the vector $\mathbf{r}$. The real number $|z|$ is called the *modulus* of the complex number z, and the angle φ is its *argument*. We shall often denote the modulus of z by ρ and the argument of z by $\arg z$, or φ (fig. 2.4). It is obvious that for the complex number $z = x + iy$,

$$x = \rho \cos \varphi$$

$$y = \rho \sin \varphi.$$

Hence,

$$z = x + iy = \rho(\cos \varphi + i \sin \varphi).$$

The expression of the number $z = x + iy$ in the form

$$z = \rho(\cos \varphi + i \sin \varphi)$$

is called the *trigonometric form* of the complex number z.

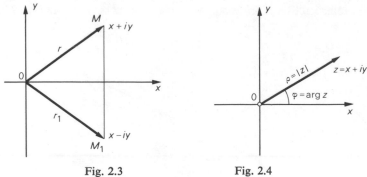

Fig. 2.3   Fig. 2.4

Along with positive angles, measured counterclockwise from the positive side of the x-axis, we introduce negative angles, which are measured clockwise from the positive side of the x-axis.

If $\bar{z}$ is the conjugate of the number

$$z = x + iy = \rho(\cos \varphi + i \sin \varphi),$$

then

$$\begin{aligned} \bar{z} &= \rho(\cos \varphi - i \sin \varphi) \\ &= \rho(\cos (-\varphi) + i \sin (-\varphi)) \\ &= \rho(\cos (2\pi - \varphi) + i \sin (2\pi - \varphi)). \end{aligned}$$

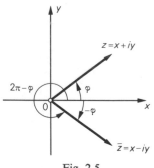

Fig. 2.5

Thus, for the argument of the number $\bar{z}$, we can take either of the angles $-\varphi$ or $2\pi - \varphi$ (fig. 2.5).

Since sine and cosine are periodic functions with period 2π, the value of the argument of a complex number z is defined up to an integral multiple of 2π. Therefore, it is convenient to select, from the values of the argument, the so-called *principal value*, which is contained within the interval from zero to 2π (inclusive of zero, but exclusive of 2π).

In the following, unless stated otherwise, we mean by the argument of a complex number z any angle φ satisfying $z = \rho(\cos \varphi + i \sin \varphi)$.

We shall now consider the multiplication of complex numbers.

Given two complex numbers $z_1 = x_1 + iy_1$ and $z_2 = x_2 + iy_2$, the product $z_1 \cdot z_2$ is defined to be the complex number

$$z = (x_1 x_2 - y_1 y_2) + i(x_1 y_2 + x_2 y_1) \,.$$

Let us consider the geometric interpretation of the operation of multiplication with the aid of the trigonometric form for complex numbers. Let

$$z_1 = \rho_1(\cos \varphi_1 + i \sin \varphi_1) \,;$$
$$z_2 = \rho_2(\cos \varphi_2 + i \sin \varphi_2) \,.$$

Then

$$\begin{aligned}
z = z_1 z_2 &= \rho_1 \rho_2((\cos \varphi_1 \cos \varphi_2 - \sin \varphi_1 \sin \varphi_2) \\
&\quad + i(\cos \varphi_1 \sin \varphi_2 + \cos \varphi_2 \sin \varphi_1)) \\
&= \rho_1 \rho_2(\cos (\varphi_1 + \varphi_2) + i \sin (\varphi_1 + \varphi_2)) \,.
\end{aligned}$$

In this manner, if the radius vector $\mathbf{r}$ represents the complex number $z = z_1 z_2$, and the radius vectors $\mathbf{r}_1$ and $\mathbf{r}_2$ represent the complex numbers z_1 and z_2, respectively, then the radius vector $\mathbf{r}$ is obtained from $\mathbf{r}_1$ and $\mathbf{r}_2$ by the following operations: The radius vector $\mathbf{r}_1$ is first rotated counterclockwise by an angle of φ_2 if $\varphi_2 > 0$, or clockwise by an angle of $-\varphi_2$ if $\varphi_2 < 0$; then, its length is increased by a factor of ρ_2. In other words, if α_{φ_2} is the rotation of the plane around the origin by an angle of φ_2, and β_{ρ_2} is the dilation transformation with coefficient ρ_2 and center at the origin, then the vector $\mathbf{r}$ is obtained from the vector $\mathbf{r}_1$ by successive application of the transformations α_{φ_2} and β_{ρ_2}. In symbols,

$$\mathbf{r} = \beta_{\rho_2}(\alpha_{\varphi_2}(\mathbf{r}_1)) \,.$$

Of course, if the roles of z_1 and z_2 are interchanged (complex multiplication is commutative), the analogous relation holds:

$$\mathbf{r} = \beta_{\rho_1}(\alpha_{\varphi_1}(\mathbf{r}_2)) \,.$$

We now turn to the geometric interpretation of the operation of dividing two complex numbers $z_1 = \rho_1(\cos \varphi_1 + i \sin \varphi_1)$ and $z_2 = \rho_2(\cos \varphi_2 + i \sin \varphi_2)$. If $z = z_1/z_2$ is the quotient of z_1 and z_2, then

$$\begin{aligned}
z = \frac{z_1 \cdot \bar{z}_2}{z_2 \cdot \bar{z}_2} &= \frac{\rho_1(\cos \varphi_1 + i \sin \varphi_1)\rho_2(\cos \varphi_2 - i \sin \varphi_2)}{\rho_2(\cos \varphi_2 + i \sin \varphi_2)\rho_2(\cos \varphi_2 - i \sin \varphi_2)} \\
&= \frac{\rho_1}{\rho_2} \cdot \frac{(\cos \varphi_1 + i \sin \varphi_1)(\cos (-\varphi_2) + i \sin (-\varphi_2))}{(\cos \varphi_2 + i \sin \varphi_2)(\cos (-\varphi_2) + i \sin (-\varphi_2))} \\
&= \frac{\rho_1}{\rho_2} [\cos (\varphi_1 - \varphi_2) + i \sin (\varphi_1 - \varphi_2)] \,.
\end{aligned}$$

Thus,

$$z = \frac{z_1}{z_2} = \frac{\rho_1}{\rho_2} \left[\cos \left(\varphi_1 - \varphi_2 \right) + i \sin \left(\varphi_1 - \varphi_2 \right) \right].$$

Let $\alpha_{-\varphi_2}$ denote the rotation of the plane about the origin corresponding to the angle $-\varphi_2$ and let β_{1/ρ_2} be the dilation transformation centered at the origin with coefficient $1/\rho_2$. Then the vector $\mathbf{r}$ is obtained from the vector $\mathbf{r}_1$ by successive application of the transformations $\alpha_{-\varphi_2}$ and β_{1/ρ_2}; that is,

$$\mathbf{r} = \beta_{1/\rho_2}[\alpha_{-\varphi_2}(\mathbf{r}_1)].$$

2.2. Linear Functions of a Complex Variable and Elementary Transformations of the Plane

Suppose every complex number $z = x + iy$ is made to correspond to some complex number $z' = x' + iy'$ by some rule. Then we say that for the set of all complex numbers, or, more simply, for the complex plane, the function of a complex variable $z' = f(z)$ is defined. A complex function whose rule of correspondence is given by the formula

$$z' \equiv f(z) = az + b,$$

where a and b are fixed complex numbers, is called a *linear* function.

Since complex numbers can be identified with points in the plane, every complex function can be considered as a transformation of the points of the plane. It is the task of the present section to describe such functions with the aid of the elementary transformations of the plane investigated in sec. 1.1.

First, let

$$f(z) = z' = az + b$$

be a given linear function. If $a = 0$, then the function $z' = b$ is constant, since it assigns the complex number b to any complex number z. The transformation of the plane corresponding to the function $f(z)$ thus takes the entire plane into the single point b.

From here on, we shall exclude this trivial transformation from our considerations and assume that $a \neq 0$.

Let

$$a = |a|(\cos \varphi + i \sin \varphi)$$

be the complex number a written in trigonometric form. Let $\mathbf{r}'$, $\mathbf{r}$, and $\mathbf{h}$ denote the radius vectors corresponding to the numbers z', z, and b respectively. Furthermore, let $\beta_{|a|}$ be the dilation transformation with center at the origin and coefficient $|a|$, and let α_φ be the rotation of the plane through an angle of φ about the origin. Finally, let γ_b be the parallel translation of the plane by the vector **h**. It is not hard to see that the point z', the endpoint of the vector **r′**, is obtained from the point z, the endpoint of the vector **r**, by successive application of the transformations α_φ, $\beta_{|a|}$, and γ_b.

A linear function of the form

$$z' = az + b$$

is often called a *linear function of the first kind*. As we have shown, a linear function of the first kind on the plane corresponds to a transformation consisting of the successive application of the transformations of rotation about the origin, dilation with center at the origin, and parallel translation. Here, the rotation and the dilation are determined by the number a, and the parallel translation by the number b.

We remark specifically on some special cases.

a. $|a| = 1$, $b = 0$: rotation of the plane about the origin through an angle equal to the argument of the number a.

b. a is a positive real number, $b = 0$: dilation transformation with center at the origin and coefficient a.

c. $a = 1$: parallel translation by the vector **h**.

The function

$$z' = a\bar{z} + b$$

is called a *linear function of the second kind*. We consider first the special case $a = 1$, $b = 0$. The function

$$z' = \bar{z}$$

takes each point z into the point $\bar{z}$ symmetric to it with respect to the x-axis. Thus, the function

$$z' = \bar{z}$$

denotes the symmetry transformation with respect to the x-axis. It is easy to see that the general linear function of the second kind corresponds to a transformation of the plane consisting of successive application of reflection across the x-axis, rotation about the origin, dilation with center at the origin, and parallel translation. Just as in the case of linear functions of the first kind, the angle of rotation is equal to the argument of the number a, the coefficient of the dilation is equal to the modulus of the number a, and the vector of the parallel translation is determined by the number b.

2.3. Linear Fractional Functions of a Complex Variable and Related Pointwise Transformations of the Plane

Functions of a complex variable given by the formulas

$$z' = \frac{az + b}{cz + d};\qquad(2.1)$$

$$z' = \frac{a\bar{z} + b}{c\bar{z} + d},\qquad(2.2)$$

where a, b, c, d are fixed complex numbers and

$$ad - bc \neq 0,$$

are called, respectively, *linear fractional functions of the first and second kind.*

We consider first functions of the form

$$z' = \frac{r^2}{z},\qquad(2.3)$$

and

$$z' = \frac{r^2}{\bar{z}},\qquad(2.4)$$

where r is some positive constant.

Equation (2.4) can be written as:

$$z' = \frac{r^2 z}{\bar{z}z} = \frac{r^2}{|z|^2}\, z\,.$$

It follows that the transformation of the plane corresponding to the function

$$z' = \frac{r^2}{\bar{z}}$$

carries the point z to the point z' lying on the ray determined by the radius vector corresponding to z, and that the modulus of the number z' is given by

$$|z'| = \frac{r^2}{|\bar{z}|} = \frac{r^2}{|z|}\,.$$

Thus z' is obtained from z by an inversion with center at the origin and radius r.

Equation (2.3) can be written as

$$(\bar{z})' = \frac{r^2}{\bar{z}}.$$

By reasoning analogous to that above, we can easily conclude that the function

$$z' = \frac{r^2}{z}$$

corresponds to the successive application of a reflection across the x-axis and an inversion with center at the origin and radius r.

We have

THEOREM 2.1. *In the complex plane, the inversion transformation T with radius r and center d is given by the function*

$$z' = \frac{r^2}{\bar{z} - \bar{d}} + d. \tag{2.5}$$

Analogously, the function

$$z' = \frac{r^2}{z - d} + d \tag{2.6}$$

yields the transformation obtained by the successive application of the reflection across a line parallel to the x-axis and passing through the point d, and the inversion with radius r and center d.

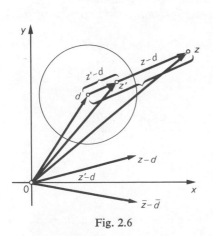

Fig. 2.6

Proof. Suppose T is the inversion with radius r and center d, and z' is the image of z under T (fig. 2.6). By the definition of the inversion T, we have

$$|z' - d| = \frac{r^2}{|z - d|} = \frac{r^2}{|\bar{z} - \bar{d}|}. \tag{2.7}$$

Furthermore, the numbers $z - d$ and $z' - d$ must have equal arguments, since the fact that z and z' lie on the same ray with initial point d implies that $z - d$

and $z' - d$ lie on the same ray with initial point at the origin. Thus, the numbers $z' - d$ and $\bar{z} - \bar{d}$ have arguments differing only in sign. Using the rule for multiplication of complex numbers in trigonometric form, we obtain

$$(z' - d)(\bar{z} - \bar{d}) = |z' - d| \cdot |\bar{z} - \bar{d}|(\cos 0 + i \sin 0)$$
$$= |z' - d| \cdot |\bar{z} - \bar{d}|.$$

This equation, along with equation (2.7), yields

$$z' - d = \frac{r^2}{\bar{z} - \bar{d}}.$$

Hence,

$$z' = \frac{r^2}{\bar{z} - \bar{d}} + d.$$

The second part of the theorem can be proved analogously.

THEOREM 2.2. *A linear fractional function of the second kind*

$$z' = \frac{a\bar{z} + b}{c\bar{z} + d}$$

with $c \neq 0$ can be written as a transformation of the complex plane consisting of the successive application of the following transformations:
   1. the inversion with center at the point $-(\bar{d}/c)$ and radius 1;
   2. rotation of the plane through an angle equal to the argument of the number $(bc - ad)/c^2$;
   3. the dilation with coefficient equal to the modulus of the number $(bc - ad)/c^2$ and center at the origin;
   4. parallel translation by the radius vector of the number $a/c + [\bar{d}(bc - ad)]/c^2\bar{c}$.

Proof. The linear fractional function (2.2) can be written as:

$$z' = \left[\frac{1}{\bar{z} + \dfrac{d}{c}} \cdot \frac{\bar{d}}{\bar{c}}\right]\frac{bc - ad}{c^2} + \left[\frac{a}{c} + \frac{\bar{d}(bc - ad)}{c^2\bar{c}}\right]. \qquad (2.8)$$

The validity of theorem 2.2 follows immediately from formula (2.8).

We have the analogous theorem for linear fractional functions of the first kind. The only difference is that between the inversion and rotation transformations, there occurs a reflection across a line passing through the point $-(d/c)$ and parallel to the x-axis.

If the coefficient c is zero for the linear fractional functions (2.1) or (2.2), they reduce to linear functions of the type considered in sec. 2.2.

3

Groups of Transformations: Euclidean and Lobachevskian Geometries

In this chapter we shall give a brief construction of the so-called *Euclidean* and *Lobachevskian* geometries from the point of view of group theory. This approach to the study of various geometries was first proposed by the German mathematician F. Klein in 1872.

3.1. The Geometry of a Group of Transformations

3.1.1. *The concept of a group.* One of the most fundamental concepts in algebra is that of a group.

Suppose G is some set, the nature of whose elements is irrelevant. For example, the elements of G may be numbers, vectors, functions, transformations, or some other objects.

Now suppose that some rule of correspondence is given under which some element c of G is assigned to each ordered pair (a, b) of elements from G. Then we say that there is an *operation* defined on G, which is normally called multiplication and denoted by a dot. That is, if the element c of G is assigned to the ordered pair (a, b), we write

$$c = a \cdot b.$$

The element c is usually called the *product* of the elements a and b. We note that it does not follow from the definition of an operation that $a \cdot b$ is always equal to $b \cdot a$.

Now suppose that an operation $\cdot$ is introduced on the set G. We say that G forms a *group* with respect to the operation $\cdot$ if the following requirements (group axioms) are satisfied:

1. The associative law: For any three elements a, b, and c in G we have the equality

$$(a \cdot b) \cdot c = a \cdot (b \cdot c).$$

2. There exists an element e in G such that for any other element a of G the equality

$$a \cdot e = a$$

holds. The element e is called a *unit element* of the group.

3. For any element a in G there exists an element x in G satisfying

$$a \cdot x = e.$$

The element x is called an *inverse* of the element a.

We shall now verify a number of simple propositions which follow directly from the definition of a group.

a. By axiom 1, no ambiguity results when we denote the group element $(a \cdot b) \cdot c$ or $a \cdot (b \cdot c)$ simply by $a \cdot b \cdot c$.

b. If e is a unit element of the group G, then for any element a of G we have

$$e \cdot a = a.$$

Furthermore, for every element a in G with inverse x, the equality

$$x \cdot a = e,$$

as well as the postulated equality

$$a \cdot x = e,$$

holds

Let us prove proposition (b). If y is an inverse of x, that is, a group element satisfying

$$x \cdot y = e,$$

then

$$x \cdot a = (x \cdot a) \cdot e = (x \cdot a) \cdot (x \cdot y)$$
$$= x \cdot (a \cdot x) \cdot y = x \cdot e \cdot y$$
$$= x \cdot y = e,$$

which establishes the second part of our assertion, along with the fact that a is an inverse of its inverse x. Furthermore,

$$e \cdot a = (a \cdot x) \cdot a = a \cdot (x \cdot a) = a \cdot e = a,$$

by what we have just proved. This completes the proof of assertion (b).

c. In the group G, each of the equations

$$a \cdot x = b \tag{3.1}$$

and

$$x \cdot a = b \tag{3.2}$$

have unique solutions in x.

It is not hard to see that if g is an inverse for a, then the elements $g \cdot b$ and $b \cdot g$ are solutions to equations (3.1) and (3.2), respectively. To show that these solutions are unique, suppose, for example, that equation (3.1) has solutions x_1 and x_2. Then, since

$$a \cdot x_1 = b = a \cdot x_2 ,$$

we have

$$x_1 = g \cdot a \cdot x_1 = g \cdot b = g \cdot a \cdot x_2 = x_2 ,$$

where g is an inverse of a. The proof of uniqueness is completely analogous for equation (3.2).

We note that by virtue of assertion (c), the unit element e and the inverse of a given element a are unique, since all unit elements are solutions to the equation $a \cdot x = a$, and all inverses of a are solutions to the equation $a \cdot x = e$. We may therefore denote the unique inverse of a by a^{-1}.

A subset H of the group G which is closed under the operation in the group G and satisfies the three group axioms with respect to that operation is called a *subgroup* of the group G. Clearly, every subgroup contains the unit element of the group and the inverse of each of its elements.

We shall now present some examples of groups.

1. The set of all integers forms a group under the operation of addition. If m is some integer, then the set of all integers of the form km, for $k = 0, \pm 1, \pm 2, \ldots$, forms a subgroup of this group.

2. The set of all nonzero real numbers forms a group under multiplication. The set of all nonzero rational numbers forms a subgroup of this group.

3. The set of all radius vectors in the plane forms a group under addition. The set of radius vectors lying on one line through the origin forms a subgroup of this group.

4. The set of all nonzero complex numbers forms a group under multiplication. The set of all complex numbers of modulus one and the set of all nonzero real numbers are two of its subgroups.

3.1.2. *The group of transformations of a set.* Let M be an arbitrary non-empty set. A rule of correspondence f which assigns an element x' $= f(x)$ of M to each element x of M is called a *transformation* of the set M into itself. The element x' is called the *image* of x under f.

The set of all images $x' = f(x)$, as x runs through M, is denoted by $f(M)$. It is obvious that $f(M)$ either coincides with M or is a proper (and non-empty) subset of M.

The transformation f of the set M into itself is called a *one-to-one transformation of M onto itself* if it satisfies the following two conditions:

1. Different elements x_1 and x_2 of the set M correspond to different images $f(x_1)$ and $f(x_2)$.

2. The set $f(M)$ coincides with the set M.

We shall consider below only one-to-one transformations of the set M onto itself, which will be referred to simply as *transformations*.

Let f be a transformation on the set M. Since $f(M) = M$, we know that for any x' in M it is possible to find a unique x in M satisfying

$$x' = f(x)$$

(the uniqueness of x arises from condition 1 above). Thus a rule of correspondence g exists which assigns to each x' in M the unique x satisfying

$$x' = f(x) ;$$

we may write

$$x = g(x') .$$

It is easy to show that g is itself a transformation and is uniquely determined by f; it is called the *inverse* of f and is denoted by f^{-1}.

Let f_1 and f_2 be two given transformations. Then the successive application of f_1 and f_2 defines a new transformation f on M given by

$$f(x) = f_2(f_1(x)) .$$

The transformation f is called the *composition* or *product* of the transformations f_1 and f_2 and is denoted $f_2 \cdot f_1$ (the transformation written on the right side of the dot is always carried out first). The composition of transformations, generally speaking, depends on the order in which they are performed; that is, in general, $f_2(f_1(x))$ need not equal $f_1(f_2(x))$.

The transformation e defined by $e(x) = x$, which leaves all elements of M fixed, is called the identity transformation. If f is a given transformation and f^{-1} is its inverse, then it is easy to see that for any x in M the relations

$$f(f^{-1}(x)) = x = e(x); \qquad f^{-1}(f(x)) = x = e(x)$$

are valid.

We have:

THEOREM 3.1. *The set of all transformations of a set M onto itself forms a group under the operation of composition.*

Verification of the group axioms in this instance is very simple.

1. If f_1, f_2, and f_3 are transformations of the set M, then

$$(f_3 \cdot f_2) \cdot f_1 = f_3 \cdot (f_2 \cdot f_1) .$$

It is easy to show that both the left and right sides of the above equation reduce to the transformation f defined by $f(x) = f_3[f_2(f_1(x))]$.

Consequently, the composition of transformations always obeys the associative law.

2. The identity transformation e plays the role of the unit element of the group. For any transformation f on M and any element x in M, we have

$$f(e(x)) = f(x) .$$

It follows that $f \cdot e = f$.

3. For any transformation f there exists a transformation g such that

$$f \cdot g = e .$$

We need only take $g = f^{-1}$.

Thus the theorem is proved.

The group of all transformations on the set M will be denoted by $G(M)$.

Any subgroup of the group $G(M)$ will be called a *group of transformations on the set M*. A nonempty subset H of $G(M)$ is a subgroup if the following two conditions hold: (1) the composition $f_2 \cdot f_1$ of any two elements f_1 and f_2 of H is contained in H; (2) the inverse f^{-1} of any element f of H is contained in H. These conditions are sufficient, since the associative law always holds on any subset of $G(M)$, and since a nonempty subset H must contain some transformation f, and thus the transformations f^{-1} and $f \cdot f^{-1} = e$ if conditions (1) and (2) are satisfied.

3.1.3. *The geometry of a group.* Let *M* be some set of arbitrary elements, and *H* a group of transformations on *M*.

In the interest of visualization, we shall call *M* a space, and its elements points. A set of points will be called a *figure*.

A figure *A* is called *equivalent* to a figure *B* if there exists a transformation *f* on the group *H* carrying *A* onto *B*.

This relationship of equivalence of figures has the following important properties:

1. *Every figure A is equivalent to itself.*

The unit element of the group *H*—the identity transformation of the set *M* onto itself—carries *A* onto *A*.

2. *If the figure A is equivalent to the figure B, then the figure B is equivalent to the figure A.*

Actually, if the figure *A* is carried onto *B* by a transformation *f* from the group *H*, then, since the inverse f^{-1} of *f* also lies in *H*, f^{-1} carries the figure *B* onto the figure *A*.

3. *If the figure A is equivalent to the figure B, and the figure B is equivalent to the figure C, then the figure A is equivalent to the figure C.*

If the transformation *f* in *H* carries *A* onto *B*, and the transformation *g* takes *B* onto *C*, then the transformation $g \cdot f$ carries *A* onto *C*; and since $g \cdot f$ lies in *H* (*H* is a group), *A* is equivalent to *C*.

By virtue of properties 1, 2, and 3, the equivalence relation divides the set of all figures into equivalence classes, with each figure lying in one and only one class.

Definition of a geometry from the point of view of group theory, as proposed by Klein, involves consideration of certain *geometric* properties and measurements of figures in a space *M* which are invariant under all transformations from a given group *H*, and are thus identical in all equivalent figures.

The set of all properties and quantities invariant under transformation by elements of a group *H* is called the *geometry* of the group *H*.

Klein's idea of regarding different geometries as sets of invariants under corresponding groups has made it possible to disclose fundamental relations among various geometries—projective, affine, Euclidean, and Lobachevskian—which were constructed and studied around 1880. The reader can find a detailed presentation of these matters in N. V. Efimov's book, *Vysshaya geometriya* [Higher geometry].

In the next two sections we shall show how the geometries of Euclid and Lobachevskii can be constructed from the point of view of group theory.

3.2. Euclidean Geometry

We shall restrict our consideration of Euclidean geometry to the plane. In sec. 1.1 we studied, in the Euclidean plane, motions which could be represented as one-to-one transformations of the plane which preserve distances between points, the so-called isometries. The corresponding sets of equivalent figures consist of those figures which can be transformed onto one another by isometries. The fact that the set of isometries is a subgroup of the group of transformations on the plane is easily verified. First, suppose that f and g are isometries. Then the transformation $h = g \cdot f$ is also an isometry. The transformation h is clearly one-to-one and onto; and if $d(X, Y)$ denotes the distance between the points X and Y in the plane,

$$d(h(A), h(B)) = d(g(f(A)), g(f(B)))$$
$$= d(f(A), f(B))$$
$$= d(A, B)$$

for any points A and B in the plane. Furthermore, if f is an isometry and f^{-1} is the inverse transformation of f, then f^{-1} is also an isometry, since for any points A and B in the plane,

$$d(A, B) = d(f(f^{-1}(A)), f(f^{-1}(B)))$$
$$= d(f^{-1}(A), f^{-1}(B)) .$$

Thus, the isometries form a group of transformations on the plane. The geometry of this group is called the Euclidean geometry of the plane.

Since any isometry (see sec. 1.1) is the composition of rotations, parallel translations, and, possibly, reflections across lines (in this connection we are allowing rotation through a zero angle and parallel translation by a zero vector, which result in the identity transformation), Euclidean geometry can be defined as a set of propositions about properties of figures and quantities which are invariant under all possible rotations, parallel translations, and reflections across lines, as well as compositions of these transformations.

In sec. 2.2, using the identification of points in the Euclidean plane with complex numbers, we showed that linear functions of a complex variable of the first and second kinds,

$$z' = az + b ; \tag{3.3}$$

$$z' = a\bar{z} + b , \tag{3.4}$$

determine one-to-one transformations of the plane, which are isometries if the modulus of the number a is one. We shall prove that, using functions of the forms (3.3) and (3.4), it is possible to specify any isometry of the plane. In fact, let f be any isometry of the plane. We may write

$$f = p \cdot g \quad \text{or} \quad f = s \cdot p \cdot g,$$

where g is a rotation through an angle α about the point $D = (d_1, d_2)$, p is a parallel translation by the vector **OB** with coordinates (b_1, b_2), and s is a reflection across the line l, passing through the point $C = (c_1, c_2)$ and making an angle γ with the positive direction of the x-axis.

The rotation g corresponds to the linear function

$$z' = G(z) = a(z - d) + d,$$

where

$$a = \cos \alpha + i \sin \alpha; \qquad d = d_1 + id_2.$$

The parallel translation p corresponds to the linear function

$$z' = P(z) = z + b,$$

where

$$b = b_1 + ib_2.$$

Finally, the reflection s across the line l corresponds to the linear function

$$z' = S(z) = u(\bar{z} - \bar{c}) + c,$$

where

$$u = \cos 2\gamma + i \sin 2\gamma; \qquad c = c_1 + ic_2; \qquad \bar{c} = c_1 - ic_2.$$

We leave it to the reader to convince himself of the validity of these facts.

Thus, the function f in the case when $f = p \cdot g$ has the form

$$z' = P(G(z)) = G(z) + b = a(z - d) + d + b = az + d + b - ad,$$

or, finally,

$$z' = az + (d + b - ad),$$

where

$$|a| = \sqrt{\cos^2 \alpha + \sin^2 \alpha} = 1.$$

If $f = s \cdot p \cdot g$, then the corresponding function has the form

$$z' = S(P(G(z))) = S(az + (d + b - ad))$$
$$= u[\overline{(az + (d + b - ad))} - \bar{c}] + c$$
$$= u(\bar{a}\bar{z} + (\bar{d} + \bar{b} - \bar{a}\bar{d}) - \bar{c}) + c,$$

or, finally,

$$z' = (u\bar{a})\bar{z} + (u(\bar{d} + \bar{b} - \bar{a}\bar{d} - \bar{c}) + c),$$

with

$$|u\bar{a}| = |\cos(2\gamma - \alpha) + i\sin(2\gamma - \alpha)| = 1.$$

From these considerations arises:

THEOREM 3.2. *There exists a one-to-one correspondence between the isometries of the Euclidean plane and linear functions of a complex variable of the first or second kind*

$$z' = az + b$$

and

$$z' = a\bar{z} + b,$$

such that $|a| = 1$; *in addition, if the isometry f is the composition of isometries* f_1 *and* f_2, *that is,*

$$f = f_2 \cdot f_1,$$

with $F(z)$ *the complex function corresponding to f,* $F_1(z)$ *the complex function corresponding to* f_1, *and* $F_2(z)$ *the complex function corresponding to* f_2, *then*

$$F(z) = F_2(F_1(z)). \tag{3.5}$$

Specifically, if

$$F_1(z) = \begin{cases} a_1 z + b_1, & \text{(3.6)} \\ a_1 \bar{z} + b_1; & \text{(3.7)} \end{cases}$$

$$F_2(z) = \begin{cases} a_2 z + b_2, & \text{(3.8)} \\ a_2 \bar{z} + b_2, & \text{(3.9)} \end{cases}$$

then, correspondingly,

$$F_2(F_1(z)) = \begin{cases} a_2 F_1(z) + b_2 = \begin{cases} a_2 a_1 z + (a_2 b_1 + b_2), \\ a_2 a_1 \bar{z} + (a_2 b_1 + b_2); \end{cases} \\ a_2 \overline{F_1(z)} + b_2 = \begin{cases} a_2 \bar{a}_1 \bar{z} + (a_2 \bar{b}_1 + b_2), \\ a_2 \bar{a}_1 z + (a_2 \bar{b}_1 + b_2). \end{cases} \end{cases} \quad (3.10)$$

The compositions of functions (3.6), (3.8) and (3.7), (3.9) yield linear functions of the first kind, while the compositions of functions (3.6), (3.9) and (3.7), (3.8) yield linear functions of the second kind. The modulus of the coefficient of z and $\bar{z}$ in all four functions, clearly, is equal to one.

The formulas (3.5)–(3.10) yield, for every pair of complex linear functions $F_1(z)$ and $F_2(z)$, a corresponding complex linear function $F(z)$ which we shall call the *composition* of the functions $F_1(z)$ and $F_2(z)$ and denote by $(F_2 \cdot F_1)(z)$.

With respect to the operation of composition, the set of linear functions of the first and second kinds forms a group. Verification of this fact is extraordinarily simple. From the very definition of composition it follows that the associative law is obeyed. Furthermore, the function $F(z) = z$, corresponding to the identity transformation on the plane, plays the role of the unit element of the group, and finally, the function

$$Q(z) = \begin{cases} \dfrac{1}{a} z - \dfrac{b}{a}, & \text{if } F(z) = ax + b, \\[2mm] \dfrac{1}{a} \bar{z} - \dfrac{b}{\bar{a}}, & \text{if } F(z) = a\bar{z} + b, \end{cases} \quad (3.11)$$

satisfies

$$(F \cdot Q)(z) = F(Q(z)) = z;$$

that is, $Q(z) = F^{-1}(z)$.

Let us consider the set of all linear functions of the first and second kinds, in which the coefficient of the variable z has modulus one. From the formulas (3.10) for the composition of linear functions and formula (3.11) for the inverse of a linear function, it follows that this set forms a subgroup of the group of linear functions introduced above. This subgroup will be denoted by E. Clearly, E is a group of transformations of the set of complex numbers.

From all of the considerations above, we obtain the following theorem:

THEOREM 3.3. *The set of invariants under the group E is the Euclidean geometry of the plane.*

3.3. Lobachevskian Geometry

In the first half of the nineteenth century the Russian mathematician N. I. Lobachevskii solved the difficult, centuries-old problem of the independence of the axiom of parallelism from the other axioms of Euclidean geometry. The new ideas developed in Lobachevskii's work exercised an enormous influence on the subsequent development of mathematics.

The system of axioms underlying Lobachevskian geometry is obtained from the system of axioms for Euclidean geometry by replacing the axiom of parallelism with a new axiom, which is a statement contrary to the Euclidean axiom. The new axiom is formulated as follows: "In any plane α containing a line a and a point A not lying on a, it is possible to pass at least two distinct lines a' and a'', having no points in common with the line a, through A."

We shall present below one of the interpretations of Lobachevskian geometry presented by the French mathematician Poincaré.

We consider some straight line l in the Euclidean plane. Without loss of generality, we can assume that the line l coincides with the x-axis. We shall call the set of all points (x, y) of the plane whose y-coordinate satisfies the inequality $y > 0$, the *upper half plane*.

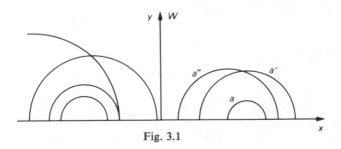

Fig. 3.1

The points of the upper half plane are taken as the points of the Lobachevskian plane. We remark that the points of the x-axis are not points of the Lobachevskian plane. Euclidean half-circles with centers on the x-axis and Euclidean rays with vertices on the x-axis which are perpendicular to the axis are regarded as lines in the Lobachevskian plane (fig. 3.1).

Two figures A and B are considered equivalent if there exists a finite number of transformations $\varphi_1, \varphi_2, \ldots, \varphi_m$, each of which is an inversion with center on the x-axis or a reflection across a line perpendicular to the x-axis, such that the transformation $f = \varphi_m \cdot \varphi_{m-1} \cdot \, \cdots \, \cdot \varphi_2 \cdot \varphi_1$ takes the figure A onto the figure B.

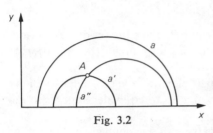

Fig. 3.2

It is evident that in Poincaré's interpretation, Lobachevskii's axiom is fulfilled (fig. 3.2). We leave it to the reader to convince himself of the validity of Lobachevskii's axiom in the cases not represented in figure 3.2.

Let W denote the upper half plane, and let H be the set of all transformations of the form

$$f = \varphi_m \cdot \varphi_{m-1} \cdot \, \cdots \, \cdot \varphi_2 \cdot \varphi_1 \quad (m \text{ is any natural number}),$$

where $\varphi_1, \ldots, \varphi_m$ are inversions with centers on the x-axis or reflections across a line perpendicular to the x-axis.

From the properties of these transformations, we already know that each of these transformations carries the upper half plane onto itself on a one-to-one basis. Consequently, the set H consists of one-to-one transformations of the upper half plane W onto itself.

We shall now prove that H is a group of transformations on the set W. If f and g are in H and

$$f = \varphi_m \cdot \varphi_{m-1} \cdots \varphi_2 \cdot \varphi_1 \, ;$$
$$g = \psi_n \cdot \psi_{n-1} \cdots \psi_2 \cdot \psi_1 \, ,$$

then, for the composition of the transformations f and g, we have the formula

$$g \cdot f = \psi_n \cdot \psi_{n-1} \cdots \psi_2 \cdot \psi_1 \cdot \varphi_m \cdot \varphi_{m-1} \cdots \varphi_2 \cdot \varphi_1 \, ,$$

from which it follows that $g \cdot f$ lies in the set H.

Since two successive iterations of the same inversion or reflection φ reduce to the identity transformation, it is obvious that

$$\varphi^{-1} = \varphi$$

and, consequently, the transformation

$$h = \varphi_1 \cdot \varphi_2 \cdot \ \cdots \ \cdot \varphi_m$$

is the inverse of the transformation

$$g = \varphi_m \cdot \varphi_{m-1} \cdot \ \cdots \ \cdot \varphi_1 .$$

The transformation h, obviously, lies in H. Thus the set of transformations H on the upper half plane forms a group under the operation of composition (see sec. 3.1.2).

The transformations of the group H play the role of isometries in the Lobachevskian plane W: they carry figures to equivalent figures in the sense of the above definition.

Therefore, Lobachevskian geometry can be defined as the set of invariants under the group of transformations H of the upper half plane W.

In conclusion, we suggest that the reader carry out the very useful exercise of formulating Lobachevskian geometry with the help of linear fractional functions of a complex variable just as was done in sec. 3.1 for Euclidean geometry.

A detailed exposition of the questions considered in chapter 3 can be found in N. V. Efimov's *Vysshaya geometriya* [Higher geometry]. A detailed presentation of Lobachevskian geometry in the Poincaré model can be found in A. S. Smogorzhevskii's book, *O geometrii Lobachevskogo* [On the geometry of Lobachevskii] (in the series "Popular Lectures in Mathematics," pamphlet 23).

1687

1637